NATURE WATCH

Alan J. Kidd
Christmas 1982.

NATURE WATCH

JULIAN PETTIFER
and
ROBIN BROWN

LONDON

MICHAEL JOSEPH

Also by Julian Pettifer (with Kenneth Hudson)

DIAMONDS IN THE SKY

Also by Robin Brown

WHEN THE WOODS BECAME THE TREES

A FOREST IS A LONG TIME GROWING

THE LURE OF THE DOLPHIN

First published in Great Britain by
Michael Joseph Limited
44 Bedford Square, London W.C.1.
1981

ISBN 0 7181 1994 0

Filmset in Great Britain by BAS Printers Limited,
Over Wallop, Hampshire
Printed and bound in Italy by Arnoldo Mondadori, Verona

Contents

Introduction

This book started out as a study of a species hitherto not examined in detail – the human naturalist.

Naturalists often go unnoticed because of the intriguing plant and animal backgrounds in which it is in their interest to conceal themselves. Having met a great number of them in the course of our work for various media we came to the conclusion that they themselves have something rather interesting to offer.

Be they amateurs or professional scientists (and there are far more amateurs than professionals), they belong to a rare community of contentment which appears to have found a direct route to fulfilment; a place where time never hangs heavy.

We further suspect that our society as a whole is seeking to share their secret. In the last decade the ecology activists have become a political force to be reckoned with; we are also entering an era of silicon technology that will produce more leisure time and require a reappraisal of the status and the meaning of the word 'work'.

Naturalists have already made this important mental adjustment. Very few of them, and certainly none in this book, would regard what they do as 'work'. In fact, the majority resent the time stolen from their natural history interests by the need to earn a living.

We would stress that the qualification 'naturalist' requires nothing more in the first instance than a real interest. Admittedly, a number of our subjects are natural history scientists, but all of them were interested in animals or plants long before they took the training that allowed them to earn livings as naturalists. An almost equal number of our subjects are complete amateurs, but they are just as eminent. Their deep interest in natural history leads to self-education and often, as our amateurs demonstrate, the ability to write books and make films which are highly respected by their scientific fellows.

And you do not need the Serengeti National Park in order to study animals, or the Matto Grosso jungle to be a plant enthusiast. Make a start with the birds and insects of your own back garden or with the lovely wild flowers of the fields. Densey Clyne had nothing but her back garden in suburban Sydney, Australia, when she began. Now she is an internationally famous insect expert and a friend to the most lethal little wriggly on Earth, the funnel-web spider.

A love of the countryside and a desire to protect wild flowers, many of which are as gravely threatened as the giant panda, will put you on a par with one of the most famous natural scientists in this book, Dr Miriam Rothschild. Dr Rothschild is a leading expert on insects, but she is as much interested in wild flowers and now spends a lot of her valuable time propagating rare species. Her aim is to produce wild flowers with seeds hardy enough to be commercially packed so that you and your children can buy and scatter them on country walks.

It is also worth mentioning that if you can read this book you have a streak of the naturalist in you by tradition, and that whole tradition was founded by amateurs. What is generally regarded as the first natural history society in the world, the Temple Coffee House Botanic Club, was founded in England almost exactly 300 years ago.

A naturalist's paradise – the Okeefenoke swamp in Georgia, USA.

In those days the only medicines were herbal ones prepared by apothecaries whose apprentices were required to serve a seven- or eight-year apprenticeship. In 1673 the Society of Apothecaries established a Physic Garden in London where the apprentices could study drug plants. And at least once a month they organized what must constitute the origin of the 'nature walk': trips into the country, called herbarizings, on which apprentices learnt to recognize useful plants growing wild.

By the time the Botanic Club was established in the Temple Coffee House 16 years after the establishment of the Physic Garden, the herbarizing walks had stopped being so specific; the science of natural history had been born and already had its pioneers.

Records show there was a flourishing entomologists' club, The Society of Aurelians, by 1740, and other clubs specializing in different areas of natural history soon followed.

The problem is that today most of us live in that monument to the protestant work ethic – the city – and although our knowledge of nature has increased academically, our contact with it has diminished. It may have diminished to such an extent as to be dangerous to us and to nature itself.

To meet our naturalists we went on a trip right round the world. But no matter which country we visited and irrespective of the plants or animals being studied, the message of alarm and, in many cases, despondency, was the same.

There is a deep-rooted fear among naturalists about what man is doing to the planet and its living organisms, a consensus of concern that hangs over this otherwise happy community like a black cloud.

Miriam Rothschild is growing her hardy wildflower strains to stem the destructive instincts of a society that cannot see it is picking and ploughing them into extinction. Bobby Tulloch, the Shetland conservationist, counts the millions of birds that depend on these islands for a vital migration-stop, knowing that the computers are forecasting a major spill of North sea oil every five years.

In Papua New Guinea, Roy Mackay watches human expansion erode the rain forests – a development that will decimate whole species. From his base in Cheyenne, ecologist Harry Harju mounts guard on the vast territory of Wyoming against the mining interests, as if he were at war.

We found no exception to this seemingly inexorable assault on nature by the naked ape, no matter what his colour or his creed. We learnt that conservationists have estimated that the final quarter of this century could witness the elimination of a million species: 40,000 a year, or about 500 by the time you have finished reading this book!

With the naturalists we met we share the view that this has unquestionably got to stop because the final death-rattle on a roller-coaster of extinction of that magnitude will be man's. Like the giant whale is dependent on one of the smallest organisms, the krill, we men who claim to sit on the mountain top of the natural kingdom only occupy that position thanks to the web of life below that supports and sustains us. Perhaps more to the point, how could the one animal on earth that claims to be altruistic, and has the intelligence that allows it to care, have created such a suicidal threat?

The last chapter records the thoughts of the world's greatest living naturalist, the Nobel prizewinner, Konrad Lorenz. Lorenz has used his extraordinary scholarship to assess man's behaviour in relation to the rest of nature and his conclusions are less than optimistic. He believes something went radically wrong with the human species at the time of the early Stone Age, when the naked ape essentially separated himself from the basic imperatives of the natural kingdom. Weapons, clothing and social organization set him free of the primary life dictates of food, warmth and the threat of predators. But he still had natural aggressive drives; these it seems were turned against other men so that for thousands of years human selection has been shaped by tribal wars leading, Lorenz believes, to 'evil selection' promoting what he calls 'warrior virtues'.

Put another way, man's distorted aggressive drive has produced a bizarre hero: the ultimately dominant man.

Also, in Lorenz's opinion, we have got ourselves trapped down a blind alley of evolution that threatens our continuing existence as a species. He points out with great logic that the 'managerial diseases' and 'torturing neuroses' produced by man's frantic career and work patterns have no real point and are, in fact, downright dangerous. He compares man to the stag which has developed huge antlers that it no longer uses for defence and which make it more vulnerable to predation.

Stags could, theoretically, selectively phase out their bizarre antlers and certainly man, with his conceptual brain, could decide to live at a quieter pace and exploit nature less. But the stags down their blind alley don't and man, who thinks he is in control of his own destiny, doesn't either.

So from our starting point of a book about a rare and ill-defined species, the human naturalist, we found ourselves dealing with something rather more important. It may well be that naturalists hold the key to our survival and by that we do not mean they are guardians of the environment, issuing vital warnings we should hear.

It is more than that. The characteristic which, initially, we, perhaps naively, identified as 'fulfilment' really describes an appropriate empathy with nature to which all men must aspire if indeed men, all other life forms and this green planet, are to survive.

In his most recent book *The Year of the Greylag Goose* Konrad Lorenz points to the solution. 'Far too much of civilized mankind today is alienated from nature. Most people seldom encounter anything but lifeless, man-made things in their daily lives and have lost the capacity to understand living things or to interact with them. That loss helps to explain why mankind as a whole exhibits such vandalism towards the living world of nature that surrounds us and makes our way of life possible. It is an important and worthy undertaking to try to restore the lost contact between human beings and the other living organisms of our planet. In the final analysis, the success or failure of such a venture will determine whether or not mankind destroys itself along with all the other living beings on Earth.'

North American squirrel from Wyoming, Harry Harju's territory.

We could pen no better introduction to this book and its purpose than that. It only remains to say that the naturalists featured in the pages to come are the elite of those trying to 'restore the lost contact between human beings and the other living organisms of our planet'. We hope you will see your way to joining their company.

MIRIAM ROTHSCHILD

Bugs, Bogs and Lady's Bedstraw

Ashton Wold, Northamptonshire, England.

Dr Miriam Rothschild is a member of the famous merchant banking family and probably the world's leading authority on bird fleas. Her interest in parasites she inherited from her father the Hon. N. Charles Rothschild but where *he* developed his inclination towards entomology remains a mystery, even to the family.

Dr Rothschild has written of her father and of her uncle Walter (Second Baron Rothschild) as being 'strange mutations': that down the years, Rothschilds had involved themselves in all manner of activities and good works but never, until their generation, in natural history. Since then however, their involvement has been on such a scale and of such quality that the name of Rothschild is as much respected in scientific circles as it is among bankers.

Dr Miriam Rothschild believes that had her father been born 50 years later he might well have adopted natural science and conservation as a career, as she has done. But in Edwardian times tradition demanded that he enter the family firm and that collecting bugs should remain a spare time activity, albeit an eccentric one. Even today the idea of a leading banker collecting fleas raises a smile; at the turn of the century it raised the roof and tended to obscure the fact that Charles Rothschild was an all-round naturalist and indeed a founder of the modern conservation movement.

Many of his tastes and interests are evident in the house and in the village, which he built in 1900, where his daughter lives and works today. On a warm afternoon in early summer Ashton, Northamptonshire is everything that an English village ought to be and so frequently is not.

Miriam Rothschild, a lifetime of staring down a microscope.

Bugs, bogs and lady's bedstraw, or, more precisely, insects, nature reserves and wild flowers are Dr Miriam Rothschild's consuming interests.

In the greenhouses at Ashton Wold, thousands of butterflies are bred for experimental purposes.

It is full of ancient peace, fresh and tranquil like the landscapes of Cotman and Constable. Architecturally it is of a piece, its fine stone-built houses and thatched roofs straggling charmingly down the lane, shaded by great trees. In the fragrant cottage gardens there is little evidence of the bungalow culture, the prefabricated garages and plaster gnomes that disfigure so much of the countryside.

When Charles Rothschild completed the rebuilding of Ashton, it became the first village in England (and probably in Europe) to have a bathroom in every house. As well as promoting hygiene, his plans aimed to foster local crafts, such as thatching and stone-masonry and to encourage gardening. All these objectives are pursued vigorously by his daughter today.

But it is the local inn that suggests the continuity of interest in natural history. It is named after a butterfly – the chequered skipper – and its bar is filled with old-fashioned lepidopterists' equipment, display cases of butterflies and an extraordinary sign that uses nails to represent the chequered skipper.

A couple of miles past the inn, in the greenhouses of Ashton Wold, Dr Miriam Rothschild is wrestling with the problem of discovering how cabbage white butterflies count – or, more scientifically – 'the assessment of egg load by *Pieris brassicae*'. It is a remarkable sight: Dr Rothschild, wearing her wholly characteristic headscarf, is moving about in a cloud of large white butterflies. The air inside the greenhouse is filled with the scent of flowers and with the fluttering of thousands of wings as the insects make their way between the vases of food plants. After many years spent studying fleas, Dr Rothschild has returned to butterflies with all the pleasure that accompanies the rediscovery of a childhood passion: she has never forgotten the thrill of holding a butterfly in her hand. Her particular favourite was the small copper 'that smart sliver of metal flying about' and the blue butterflies which she thought of as 'a piece of sky fallen to earth'.

Miriam Rothschild began collecting butterflies for her father when she was five years old, but as she grew older and learned that her uncle Walter had the largest collection of butterflies ever assembled by one man, with $2\frac{1}{2}$ million specimens – 'all arranged in drawers like captive angels' – she determined to follow a different path. 'I felt that everything there was to know in the world of butterflies was already known to my uncle . . .

so I took up the study of fleas, which was my father's line really. He discovered the plague flea in Egypt, the one that carried bubonic plague, and described it in 1903.'

For half a century Miriam Rothschild continued her father's work, but now in her 70s she has abandoned fleas and returned to butterflies; 'a second childhood', she calls it, 'but you can see them even when your eyesight is failing . . . if you stare down a brass tube for 50 years, which is what I did with a miscroscope, you want to get back to something you can really see flying about.'

Dr Rothschild has long been interested in the inter-relationship between plants and insects: 'they've come down the ages together; the plants need the butterflies for pollination and the plants can provide in return a bit of foliage for the caterpillars'. She observed that insects avoid putting too many eggs on one plant so steering clear of the danger that the caterpillars will literally 'eat themselves out of house and home'. But how does the cabbage white butterfly which she is studying do its sums and avoid overloading a particular cabbage? Dr Rothschild designed a number of simple and ingenious experiments which have revealed some of the answers.

To find out if the butterflies relied on visual information, cabbage leaves were decorated with sham egg clusters composed of tiny coloured beads, bits of coloured sugar or blobs of yellow paint. She discovered that the butterflies were partially deterred by the fake eggs, but as large white butterflies mostly lay their eggs on the undersides of the leaves

How do cabbage white butterflies count? Dr Rothschild demonstrates how artificial egg-clusters have revealed some of the answers.

where an insect in flight would not readily see them, the deterrent signals were not only visual. Dr Rothschild therefore guessed that the eggs must give off a repellent scent which deters the butterfly from laying, and she was able to demonstrate this conclusively with another simple experiment. Two equally attractive looking cabbage leaves were presented to the butterflies. One, however, had been smeared with a 'soup' made from their squashed eggs. Within 10 minutes the clean leaf (the control) had received five egg batches while on the leaf treated with egg fluid, not a single egg was laid. According to Dr Rothschild 'that's the perfect experiment. You don't often get one of those'.

Further experiments have indicated that the butterfly also receives messages about the condition of the leaf and the number of eggs laid upon it through the sensory receptors on its feet and through the tip of its abdomen. But the evidence so far suggests that it is the pheromones – the repellent substances associated with the eggs – that are most important in discouraging the laying female, and Dr Rothschild is now working to discover more about these chemical messengers.

As well as the cabbage whites, there are a number of monarch butterflies flying about the greenhouse and, in Dr Rothschild's view, their relationship to plants is even more interesting. For the monarch butterfly a plant gives protection: the caterpillar feeds on leaves of certain plants containing cardiac glycosides (heart poisons), which it stores and retains until it becomes a butterfly. Being highly disagreeable to birds, these substances protect the monarch against attack 'because once a bird has eaten a butterfly containing those poisons, it will certainly never repeat the experience'.

The evolution of plants and butterflies reveals a complex pattern of give-and-take. The plants need the butterflies for pollination and provide, in return, food for the caterpillars.

Some years ago, Dr Rothschild and her co-workers were able to prove that a wide range of insects – moths, beetles and grasshoppers as well as butterflies – do protect themselves by setting apart and storing toxins taken from their food plants. Moreover, almost all the insects thus protected have a very gaudy colouration which is seen as an evolved warning system. Birds have superb vision and once they recognize a pattern that spells danger, Miriam Rothschild is convinced that they never forget it. To make matters even more interesting, other butterflies without the protective poisons have developed a striking similarity to their unpalatable neighbours, which successfully fools the birds into leaving them alone. These so-called 'Batesian mimics' Dr Rothschild more colourfully describes as 'sheep in wolves' clothing' and she has frequently observed how effective their disguise can be.

In a nearby greenhouse she keeps three tawny owls to do her 'insect sorting'. Owls, of course, have magnificent night vision and by offering them a selection of moths, toxic and non-toxic, warningly coloured and cryptically marked, she can test the ability of the insect to outwit the bird and that of the bird to spot and remember the unpalatable insect. She has even treated innocuous moths with poisons and found that the birds learn very rapidly to recognize and avoid them.

Recently, however, Dr Rothschild's favourite owl, Pamela, has become temperamental 'She is a very good bird for sorting nocturnal insects, but I have to wear protective goggles when I go to see her because she has become neurotic'. A few weeks before Pamela laid a phantom egg, 'it's a sort of false pregnancy with birds; and having laid this phantom egg she brooded it for three weeks and was perfectly happy . . . but then one of the other owls here pinched her phantom egg, occupied the nesting box and started to brood it instead. Since then Pamela has become deranged; she's so neurotic she's moulted all her feathers and given up hooting – and she's not very nice to me.'

Insect sorting at Ashton Wold, using crows and tits, has shown how very quickly day-flying birds can pick out the patterns and colours associated with poisonous insects and how long they can retain the information. One of the crows remembered a specific poisonous pattern for over a year. Dr Rothschild is also convinced that there is a good deal of teaching between birds, and that information on insect

Despite her breakdown, Pamela's night vision is still effective in detecting poisonous moths.

recognition is passed from adults to their young. And it is just as well: some butterflies and moths are so toxic that if they are swallowed by birds they cause severe vomiting, distress and even death. One African moth, which harvests its poison from a type of vetch, is so lethal that it can even kill horses and cattle if it is swallowed accidentally.

But according to Dr Rothschild, if they feed regularly on insects, mammals also have the ability to identify the dangerous species. In a small dressing room adjoining her bathroom lives Victor the fox cub, who spends much of his time sunning himself in a wired-in run on the lawn outside. Dr Rothschild is very anxious to point out that the fox is not a pet. 'This is where conservation and scientific investigation meet; I have an animal here which is wild. Its mother was killed so I brought it here to look after it and it will eventually be released. But in the meantime I'm using it for scientific investigation. In the course of my experiments I want to find out which insects are toxic and which are evil tasting, so I'm feeding the cub with insects which it will eat or reject depending on whether they're nice or nasty.' Despite what she says the manner in which she talks to the fox is much more that of a doting parent than of a dispassionate scientific investigator. It may not be treated as a pet but it is certainly not treated like a laboratory animal, an experimental tool to be used and discarded.

Victor, the orphan fox-cub, demonstrates his way of sorting insects.

Foxes are useful for Dr Rothschild's experiments because insects form such an important item of their diet; and since they depend so heavily on taste and smell, rather than on vision, she can test another angle of the insect's defence system. Some foxes can distinguish by smell between a toxic moth and one that is harmless.

One such toxic moth – the burnet – Dr Rothschild has in her private nature reserve. 'It's got every kind of protection including a totally unknown poison used to protect the eggs. It's very dangerous indeed.' In an overgrown meadow at the edge of the reserve, Dr Rothschild is in ecstasies over the burnets. The 'exquisite red and burnished green' moth excites her for two reasons: 'it's aesthetically very, very pleasing and secondly it's the most poisonous moth so far discovered'. Dr Rothschild has proved this to herself in the most convincing way: 'About 20 years ago, long before anyone knew that moths were toxic, I wanted to prove my belief that they were. If you touch them a little drop of yellow blood comes out of their heads, and I was stupid enough to scratch my hand and rub some of it in. I really thought I was going to die: I got the most terrible palpitations of the heart which must have revved up to about 150 beats a minute; I had constriction of the blood vessels and went ashy pale. I kept walking up and down saying "God, what a bloody fool, to kill yourself with a moth". The doctor came and, of course, had no idea what the antidote was. Then after about an hour I was quite

The burnet moth, successfully re-introduced into the nature reserve at Ashton Wold, is admired for its beauty and its chemical defenses. As Dr Rothschild knows to her cost, the burnet is a most poisonous moth.

normal again; but it taught me a lesson.'

It is a lesson that is learned equally well by birds. Dr Rothschild recalls how some tits reacted after they were unwise enough to consume burnets: 'I should think they had an awful pain they never forgot, because they didn't see another burnet for a whole year and when they did they wouldn't look at it. If we put it in a dish with succulent insects they love to eat they spilled the whole dish.'

One of the poisons found in the eggs, the caterpillar and the moth itself is prussic acid. At first Dr Rothschild thought that the moth absorbed this substance from one of its food plants 'but this is a snare to the biologist because if you rear the moth on another plant that doesn't contain prussic acid, the moth still has the same amount; so the insect makes the poison itself. If you squash it in your fingers you can smell it; it's like bitter almonds, very easily distinguished by a smart animal like a fox.'

Burnets are fascinating in other ways too. Dr Rothschild has noticed that males can detect the female's cocoon 'and will wait there until she comes out and then instantly copulate with her. With some South American butterflies the male actually copulates through the wall of the chrysalis.'

It seems that in the insect world – or at least among the lepidoptera – it is not unusual for the male to take advantage of the helpless female in this way. Consider the male monarch butterfly: according to Dr Rothschild 'what he does to the female is straightforward rape. It's the best example of the male chauvinist pig that you can possibly imagine. More often than not he knocks down his female and while she's half dazed has his way with her.' Worse still, Dr Rothschild has observed cases where another male will attempt to join in and mate promiscuously with either of the coupling butterflies and sometimes with the member of his own sex.

Back in the greenhouse among the cabbage whites, she reveals that *their* behaviour is by no means exemplary. The males are inclined to take advantage of any female that happens to become trapped by the araujia, or butterfly plant. To encourage pollination this beautiful bell-like flower is able to snare a butterfly by the tongue while it is searching for nectar. In its efforts to free itself the butterfly becomes covered in pollen which it will take to the next plant when it escapes – if indeed it does escape! 'Sometimes the mechanism doesn't work and you can see the white butterfly hanging by its tongue, caught in the trap; and if it happens to be a female, then the males come along and rape it. I've even seen them rape a corpse; they are violent creatures.'

If the monarch is the supreme example of the male chauvinist, the other members of its genus (Danaidea) must be regarded as the prince charmings of the butterfly world. Their sexual behaviour is of great interest as it provides a superb example of the close-knit relationship between insects and plants.

The male danaids collect certain chemicals from plants which they use to manufacture an aphrodisiac or 'love dust'. When the males are excited by the presence of a female, they flutter in front of her and extend hair pencils – 'feather dusters which they push out at the end of their bodies'. Using the hair pencils, they shower the female with a romantic cloud of love dust, which stops her in her tracks. 'She is so completely bemused by the scent that comes from these hair pencils, that she forgets to fly off and allows him to have his way with her'. What the butterflies are doing with their plant extracts is very much like what we are doing with flower extracts – manufacturing a perfume to act as an aphrodisiac; something to think about the next time you buy a bottle of scent!

If the ardent human male has little to learn from the butterfly about seduction, another area of investigation at Ashton Wold may have valuable lessons of a different kind. Hidden in an obscure part of the grounds is a greenhouse containing a crop we would not normally associate with a distinguished septuagenarian scientist. Nevertheless, there is no mistaking Dr Rothschild's invitation to look at her cannabis. Regulations for growing this plant are very strictly enforced, and the plants are carefully counted by the authorities to make sure that they are being used for their legal purpose: to feed caterpillars, not people. The objective of Dr Rothschild's cannabis cultivation was to find a good biological test for its different varieties, a test that would show that different strains have different properties and in this it has been spectacularly successful.

The caterpillars of the garden tiger moth (*Arctia caja*) have an appetite for poisonous food plants and the ability to store the plants' toxins in their bodies. One batch of these caterpillars was fed on Turkish cannabis, another on Mexican cannabis and a third control batch, on ragwort. Mexican cannabis

In the greenhouses vases of food plants for butterflies fill the air with the scent of flowers.

is known to be rich in a psychoactive substance, what Dr Rothschild calls 'the thing that gives you the big high', and the tests have proved it to be much more toxic than Turkish cannabis. 'The caterpillars ate it readily enough, but about 90 per cent of them died and those that survived grew very, very slowly. Instead of taking about six weeks to turn into butterflies, they took about nine months. So if you want to live for ever all you've got to do is to feed exclusively on cannabis.' Not that Dr Rothschild would recommend such a diet: 'After my work with moths on cannabis, I wouldn't care to smoke it because it is certainly very toxic indeed – at least to caterpillars.' The experiments also showed that since the caterpillars stored the cannabis toxins in their skins, they were able to dispose of them at each moult. But accumulating the poisonous substances between moults made an excellent predator deterrent.

Charles Darwin, who in his youth was obsessed with hunting and shooting, later put on record his discovery that the pleasure of observing and reasoning was much greater than that derived from sport. Miriam Rothschild takes this further: she describes scientific research as 'the image of war without its guilt, all excitement and challenge'; and she deplores the ultra-academic approach, the stuffiness that so frequently makes research seem dull and pedestrian.

A few years ago she wrote an article for the periodical *Antenna* entitled 'Need we be such unmitigated bores?', in which she takes her fellow scientists to task for their failure to write readable, let alone polished, English. She is even harder on the editors of learned journals who seem to her determined 'to edit out every touch of individuality from scientific papers and produce something equivalent to that variety of homogenized milk served up in American hotels'. She laments the fact that today 'abuse, as well as individuality has been expunged from formal scientific writing, and with it much of the liveliness of the proceedings'. Above all, she deplores the fact that 'sentiment – and the profound love of nature – which characterized the rather naive and ebullient entomologist of the Victorian era is also banned. It is too embarrassing.'

Not too embarrassing, fortunately, for Miriam Rothschild. Perhaps she is sufficiently eminent to be able to get away with it; but however abstruse and serious and complex the subject, her own amazement at the simple beauty of scientific discovery is always breaking through. And so is the humour: 'On one occasion I tried to enliven a particularly dreary paragraph on museum mislabelling by citing the substitution of fornicarium for formicarium. The editor was not amused.' The pleasure she gets from observing the curious sexual activities of the insect world must already be evident, but she can be just as startled and stimulated by staring down a microscope at a slide of the eighth sternite and terminal portion of *Orneacus rothschildi* (a British bird flea).

In an article entitled 'Hooked on Histology' she explains how she called her children's attention to the magic of the world of cells by projecting onto a wall, 35 mm transparencies

of cross sections of bone, gut and of squamous epithelium, trusting that the design and drama of the pictures would help them to learn. Her scheme worked better than she could ever have expected. Not only did the children pass their exams but in the process she herself 'had been irrevocably recaptured and refascinated by the sheer beauty of histology. I had to go on. I don't pretend to understand the strange aesthetic satisfaction and somewhat infantile excitement gained by recording these dreamlike, almost surrealist, landscapes of purple cells, violet and scarlet proteins, sky blue and golden cuticle. This is a mescaline-like world of fantastic colours and timeless beauty. At least I am sufficiently honest not to delude myself: this is not science, it is pure escapism and undiluted pleasure.'

She is not the kind of person to brood over regrets. And yet when she talks of her uncle Walter Rothschild, it is clear that she envies him the opportunity for scientific exploration and discovery that came to his generation and which he seized with such gusto. Lord Rothschild organized his assault on natural history like a military campaign. He lived at a time when the world's fauna was still unknown; he had 400 collectors searching the various oceanic islands for new species; he wrote about 800 scientific papers and described 5,000 species of animals. 'In a way it is less exciting today;' Miriam Rothschild admits 'it must have been tremendously thrilling to go off into the Congo and discover the okapi . . .'

Dr Rothschild delights in her uncle's eccentricity as much as in his scientific

The kitchen garden, now a riot of wild flowers, where once onions and carrots grew in regimented rows.

achievements. It seems that he was particularly fond of zebras: so much so that he had a zebra team broken and harnessed to a carriage in which he was conveyed around London. His brother Charles remarked that the zebras' camouflage was so effective that 'when they were half way down Piccadilly they vanished from sight and the cab bowled on by itself'. Dr Rothschild recalls the occasion when Queen Alexandra, who had heard of the zebra conveyance, commanded that it appear at Buckingham Palace for her inspection: 'which Lord Rothschild actually did; he drove his four-in-hand into the courtyard of the palace. He was absolutely petrified because the Queen would come up and try to pat the animals, and of course they'd kick the eye out of a gnat, those zebras. They were so difficult to harness that at Tring [the family seat] they used to approach them from above and let the harness down from the ceiling.'

And not only zebras were put to use by Lord Rothschild: his niece has a photograph of him astride a giant tortoise he had collected in the Galapagos Islands: 'it bore his weight too and he was an enormous man; he weighed 22 stone stripped and had a huge head. He was wild about these tortoises and wrote a very important book about them. He was fond of everything outsize – in fact, he was a megalomanic about size.'

It was Walter Rothschild's liking for the heavyweights of nature that led him to include cassowaries in his bird collection at Tring 'but they were banned after one of them chased the first Lord Rothschild'. The thought of the old gentleman pursued by a large irate bird obviously appeals to Miriam Rothschild; laughter comes easily to her.

Within the grounds at Ashton Wold is the 500-acre nature reserve established by Charles

Lord Walter Rothschild achieved remarkable success in taming and training a team of zebras. He also assembled the finest private collection of zoological specimens in the world. His niece, Miriam Rothschild (right), has published more than 250 scientific papers.

Rothschild at the turn of the century. A part of this reserve has *never* been ploughed: it is one of those great rarities in Britain – a stretch of virgin forest. It is Dr Rothschild's view that her father's most important contribution to science was his promotion of conservation: 'he more or less invented it in its modern sense'. He founded, and to a large extent funded, the Society for the Promotion of Nature Reserves, a ginger group of activists that vigorously promoted the revolutionary idea that the preservation of habitat, rather than the protection of individual rare species, was what really mattered. Although all this is now accepted wisdom, Charles Rothschild seems to have been given little credit for his foresight and imagination. At least the value of the nature reserve he established has been warmly appreciated by his daughter.

But the gardens are not what they were. With fuel at its present price the hothouses are no longer heated, and instead of exotic tropical plants, they shelter beds of cabbages for the butterflies. It is important for the egg-laying experiments to have a supply of cabbage leaves that are as far as possible identical. Dr Rothschild explains: 'The butterflies' choice of leaf must not be affected by any difference in the quality of the leaves, and butterflies are very choosy about where they lay. The leaves must come from the same positions on the plant or the level of mustard oils will be different; they must be the same size and colour and not blemished in any way. It's amazing that in order to get an even quality of leaves you have to grow a tremen-

dous number of cabbages: a rather boring crop but important.'

The scores of gardeners who worked the great walled garden behind the greenhouses in her father's day would be horrified if they could see it now. It is filled with poppies, harebells, yarrow, common toadflax, ox-eye daisies and lady's bedstraw: weeds that would have been ruthlessly removed and carted off to the bonfire. Today they are growing in neat rows like onions or carrots, each bed carefully labelled: sweet violets, harebells, lady's smock, primroses and row after row of cowslips. 'It was a sight for the gods when they were all in bloom together, and something that's never been seen before, a field of cowslips.'

This wild flower project is very close to Dr Rothschild's heart. Like many naturalists she is deeply concerned about the loss of so many wild plants as a result of improved agriculture, drainage of fields and the use of herbicides. 'Cowslips are getting more and more rare, and I want people to grow them in their gardens and highway authorities to grow them along motorways'. Her cultivated cowslips are much larger than those in the wild. 'When they grow in a garden without competition we get many more seeds; in five rows here we may have 4½ lb of seed, enough to plant this entire garden with cowslips.'

Although the growing of wild flowers from seed is still in an experimental stage, seed from Ashton Wold is already being distributed through commercial channels. Where, doubtless, there was once a carefully tended herbaceous border is now a wild tangle of lady's bedstraw, wild carrot, poppy and thistle. Even the thistle is welcomed 'within reason, for the butterflies.'

Dr Rothschild believes passionately that in a country the size of Britain, with its intensive agriculture and urbanization, it is not enough to have nature reserves; rather 'we have to try to cultivate plants off-site as well as on-site, and of course gardens are the obvious place. Everyone has to get into the act and help conserve our wild flora and fauna, because it's irreplaceable: every plant and every animal is the end product of millions of years of evolution and intense selection, and I don't think mankind will ever be able to reproduce the pressures that produced these perfect organisms.

'If you'd said to anyone 100 years ago "your great grandson will owe his life to some miserable little mould like penicillin" they simply wouldn't have believed you. Furthermore, one of the commonest butterflies we have around here is a source of new antibiotics. The facts haven't been published yet but it's an open secret among a certain group of scientists that about 100 new antibiotics have been found in this very common butterfly.' Clearly, if the food plant of that butterfly had been wiped out and the insect had become extinct, then man would have been the loser. Dr Rothschild points out that all garden vegetables and food crops have been selectively bred from wild plants and that we need 'wild genes' to produce new and better strains. She believes passionately that even the humblest weed is worth conserving, that man's first priority must be to maintain the rich and vital variety of living organisms, and that 'it's terribly important to preserve that gene pool'.

That is the voice of the scientist; but in any conversation with Miriam Rothschild the voice of the romantic, of the artist, of the lover of literature, always intrudes. In one of her works on parasitology *Fleas, Flukes and*

Cuckoos, there are no less than 53 quotations from sources as varied as Aristotle, Cervantes, Browning and the bible. In one breath she speaks of the practical implications of conservation in preserving the gene pool, and in the next she is musing on the mystical quality of the colour green: 'If you look at the writings of great men like Goethe and Marcel Proust, they realized that green has a special significance. Man evolved during millions of years always in green surroundings, but now in the modern world we are surrounded by concrete. The significance of this has not been thoroughly investigated. I think conservation in its widest sense brings a peace of mind which is very difficult to define – just walking under trees, among greenery gives you something that you cannot get anywhere else.'

Dr Miriam Rothschild is, at present, working on five books including a biography of her uncle Walter and a study of butterflies in art and literature. From her pen comes a constant flow of scientific papers (250 so far), articles and monographs. In the scientific world and among conservationists she is immensely respected; and her work continues. Sitting on her terrace among the wild flowers, dressed in her highly individual clothes and, as always, in rubber boots, she makes plans for the future. She recalls that many years ago, when her interest was in marine biology she studied a marine snail that grew to a gigantic size when it was castrated by one of its parasites. 'It was in the interests of the parasite to have a very large thing to eat and the snails get larger because they're castrated.

'But I believe that these parasites produce a growth hormone, and that's one of those things, in my old age, I'm going to look at.'

Reference

ROTHSCHILD, M. and CLAY, T., *Fleas, Flukes and Cuckoos. A Study of Bird Parasites*, Collins, 1952.

BOBBY TULLOCH

Bad News from Birdland

Mid Yell, The Shetland Isles, UK

Some places in the world seem much better suited to wild animals and birds than to men. One such is the most northerly part of Britain – the Shetland Isles.

Birds by the thousands, seals, whales, dolphins and a number of small land mammals have, until fairly recently, enjoyed the wild and windswept tranquillity of this seascape with virtually no human interference. But within this environment is one man charged exclusively with the well being of wildlife, and birds in particular – Bobby Tulloch, Officer for Shetland, of the Royal Society for the Protection of Birds.

The Shetland bird population, especially if you count seasonal visitors, is quite phenomenal and probably unique. It is composed of three main groups of birds: the breeding residents, the migrants that spend some time there and the passage migrants.

'For the birds of passage we are a filling station – a refuelling stop in the middle of the North Sea. They use us in emergencies only. Ironically for the bird watchers, if the weather is good the birds of passage don't stop: they make it to the mainland in one hop. But if the weather is bad, the birds come down in huge numbers. What is good news for bird watchers is bad news for the birds! If they did not have us as a refuelling stop, I have no doubt that thousands would die.'

Shetland sunset. Clear air and tranquil waters provide a perfect background for the many species of sea birds that find sanctuary in this most northern extension of the British Isles.

The variety of birds on the islands is staggering. In *A Guide to Shetland Birds*, a book co-authored with the late Fred Hunter, Tulloch listed more than 100 species that had used, or were using the island for breeding.

Bobby Tulloch looks after Shetland for the Royal Society for the Protection of Birds.

Common seals and other marine mammals also enjoy the peace of remote Shetland islands.

The list of regular or fairly regular migrants and winter visitors is also impressive.

The attraction of the Shetland Isles for birds lies in their geographical position. A group of what might be called 'temperate' birds find the place quite cold enough: they breed there in summer and fly south for the winter. Their places are then taken by a more hardy group, which has enjoyed the Arctic summer but finds it more comfortable to winter in Shetland.

Every spring and autumn Shetland is like a seaside resort for birds. There is hardly time to make up the beds before the new guests arrive. This situation becomes a little difficult to imagine when you realize that nowhere on the islands is there a single indigenous tree! But although Bobby Tulloch admits that we tend to associate inland birds with trees, they are not as important as we may think: 'Here we have crows that nest on telegraph poles and starlings that nest in holes in the ground.'

Tulloch is continually patrolling the islands, plotting bird populations, noting the number of dead birds on the beaches and keeping a somewhat fatherly eye on the Shetland mammals: a number of otters and a large number of grey and common seals. Physically he fits well into his chunky, clinker-built fishing boat: thick-set, bearded, bright eyed and, on the subject of animals,

BIRDS KNOWN TO BREED IN THE SHETLAND ISLES

Red-throated diver
Little grebe
Fulmar
Manx shearwater
Storm petrel
Leach's petrel
Gannet
Cormorant
Shag
Grey heron
Mallard
Teal
Widgeon
Pintail
Shoveler duck
Tufted duck
Long-tailed duck
Velvet scoter
Common scoter
Eider duck
Red-breasted merganser
Shelduck
Mute swan
Whooper swan
White-tailed eagle
Peregrine falcon
Merlin
Red grouse
Quail
Pheasant
Corncrake
Moorhen
Coot
Oystercatcher
Lapwing
Ringed plover
Golden plover
Snipe
Woodcock
Curlew
Whimbrew
Black-tailed godwit
Common sandpiper
Redshank
Greenshank
Dunlin
Red phalarope
Great skua
Arctic skua
Great black-backed gull
Herring gull
Common gull
Glaucous gull
Black-headed gull
Kittiwake
Common tern
Arctic tern
Sandwich tern
Razorbill
Guillemot
Black guillemot
Puffin
Rock dove
Wood pigeon
Collared dove
Cuckoo
Snowy owl
Long-eared owl
Skylark
Swallow
House martin
Raven
Hooded crow
Rook
Jackdaw
Wren
Fieldfare
Song thrush
Redwing
Ring ousel
Blackbird
Wheatear
Stonechat
Reed warbler
Blackcap
Whitethroat
Willow warbler
Goldcrest
Meadow pippit
Rock pippit
White-pied wagtail
Starling
Twite
Chaffinch
Corn bunting
Reed bunting
House sparrow
Tree sparrow

talkative, he is very proud of these islands.

His grandfather was born on one of the 90 small islands that were inhabited until the turn of the century. He can trace his island ancestry back 600 years to the Norsemen. As a Shetlander, Bobby has the consolation of knowing that the deserted settlements have become the most ideal habitat for birds.

There are nearly 100 uninhabited islands where birds and animals can 'get a bit of peace and quiet'. Only the occasional shepherd or fisherman in a boat has any reason for visiting them. A typical island with a good variety of habitat, especially low shingle beaches, will

Whereas elsewhere in Britain otters are now rarely seen, on Shetland several hundreds are flourishing.

The gregarious guillemots whose young will launch themselves over 200-foot drops, out of reach of predatory skuas, when they are only a few days old.

attract shore birds such as oyster catchers and ring plovers. If there is a bit of grass you can find curlews breeding, while further up the hill in the heather and damp ground are dunlin. In the little lochs the red-throated divers will be feeding their young, while there is usually a skua colony or two. 'The little islands are fine places for wildlife. Seals lay up around the shores, otters live in the earthy banks; all are in peace.'

Shetland lies in the middle of a shallow and very rich area of the sea. With its very long coastline – at least 1,250 miles because of all its indentations – it offers birds a very wide choice of nesting places. 'The high cliffs, with their inaccessible ledges, are favoured by birds such as guillemots which will sit in rows, jam-packed along the ledges. Razorbills are to be found down among the boulder tangles, which they share with the shags at the base of the cliffs. And where there are grassy slopes and any form of holes you find the puffin colonies – often in their thousands. The puffin is probably one of the best known and best loved of the sea birds, and in some places around the west coast of Britain their colonies have been declining seriously over the years. But we seem to be lucky in Shetland, our puffins are maintaining their numbers. On the headland of Herma Ness alone there must be more than 100,000. Places like Herma Ness are tremendously exciting – cities of birds.'

The puffin has always been one of Bobby Tulloch's favourite birds. Once, as a young child, he took a dead one home to keep as a doll!

June is the peak of the breeding season, when the birds are feeding their young. In July on the high ledges, 200 feet above the sea, in the middle of the night, young guillemots only 8 or 10 days old are leaping into the void: 'Little fluffy chicks with no wings other than little stumps jump off these ledges and land in the sea, presumably to avoid predators like the black-backed gulls and the skuas.'

An indication of how well birds can do in Shetland is the encouraging story of the great skua, or 'bonxie'. Once described by the author Eric Linklater as 'a large dark-brown, heavily built rascal of a gull that lives by piracy and will not hesitate to attack a gannet', Bobby Tulloch knows that bonxies will sometimes not hesitate to attack people if they walk through their colonies. 'I've been hit a couple of times. And if you get hit by a couple of pounds of bonxie you know they mean it!' But he has admiration for what he thinks is a courageous bird and regards the present bonxie population as a grand testimonial of the islands as a bird habitat and of the work of contemporary conservationists. 'It's a survivor! I have a sneaking respect for something that has the ability to be flexible enough to take advantage of all sorts of situations. The same goes for hooded crows. I must admit I like crows. They can survive in the face of whatever persecution comes their way. I used to keep tame crows when I was a child. A favourite ploy amongst Shetland youngsters at that time was to take a young crow from a nest to rear – but never to keep in captivity. Although some of the crows flew away when they grew big enough, most of them stayed around because they knew what was good for them; but they became a real menace with their thievery. I used to get into terrible trouble because of the things my crows did!'

Bonxies were also raised by young Shetlanders in Bobby's youth, a situation that would have been impossible 100 years before. The islands now have 96 per cent of the British great skua population – almost the entire population of the northern hemisphere. A century ago they were down to four pairs on Herma Ness and the island of Foula. 'This was one of the first practical conservation attempts on Shetland. When the bonxies on Herma Ness looked as though they were finished – because they were such rare birds they had suffered the attentions of the early egg collectors – the then owner of Herma Ness employed a man to look after the last few pairs on the headland; they never looked back. Now Foula also has approaching 3,000 pairs. That's a lot of birds.'

And that is the reason why contemporary bonxies have such a bad name; they are heavy predators on other sea birds, attacking and sometimes killing gulls larger than themselves, such as black-backs.

Like the great skua, the elegant gannet has found an ideal home in Shetland. None bred there at all until 1911, when a small colony started near Lerwick and then spread to much-favoured Herma Ness, where today there are between 5,000 and 6,000 breeding pairs. The Isle of Noss has a similar colony. 'The gannet really is a splendid bird. Its feeding behaviour – a plunge dive from a great height – is a splendid thing to watch. In late summer, the sand eels shoal, providing food for gannets. The bird will dive down into a sand eel shoal and literally gorge itself until it can hardly fly. But they do have to travel a long way to feed and then carry it back to the young, 20 or more miles away on Noss or Herma Ness. Of course the bonxies living on the islands on the way, know this. It's interesting to note that a bonxie will never attack or even look at a gannet flying away from a colony because it knows it's got nothing. They attack the gannets flying home. Always attacking the last in line, they use various tricks to force it down onto the sea and then annoy it until the gannet throws up its crop of food allowing the bonxie to help itself.'

After a lifetime observing sea birds, Bobby Tulloch no longer makes value judgements about the behaviour of particular species. He

The first gannets bred on Shetland in 1911. Now there are several thousand breeding pairs courting and nesting on the rocks.

Shetland now has 96 per cent of Britain's great skua, or bonxie, population when once they were down to a few pairs.

knows that gannets themselves are great thieves, especially during the nest building season when they continually steal material from each other.

Such behaviour can sometimes get the birds into trouble. 'Fishermen throw away a lot of pieces of terylene and polythene, which seem to have an irresistible appeal to the gannets. I've seen one bring some of this home, get in a fight over it with a neighbour until it became tangled up in it and died.'

Often wondering why gulls live in such noisy crowds, Tulloch concludes that they have learned there is safety in numbers. 'Also if they go out in parties and one finds a shoal of fish, they can all take advantage; if they are all living in the same place warnings can be communicated. And I don't doubt that they can communicate with each other in ways that we don't even begin to understand.'

But success is not always dependent upon a protective colony. The fulmar, a pair of which first came to the Shetland Isles in 1876, is now their most numerous breeding bird. 'It's a beautiful bird to watch, a marvellous glider. It's said that the first pair came to Shetland feeding on a dead whale. That pair nested on Foula. Again they seemed to find it to their liking because they never looked back and now they are everywhere: the high cliffs, the low shores, inland, even on roofs of occupied buildings. It's about the only problem I have in my garden: they fight on the garden wall, nudge each other off and being gliding birds,

Fulmars first came to Shetland 100 years ago but are now the islands' most numerous seabird.

find it difficult to take off again. I have to put them out of the garden.

'If you have ever tried handling a fulmar you will know what that involves. They have a rather revolting habit of emptying their stomach contents over anything or anybody that comes close. It is revolting – it smells to high heaven and doesn't leave your clothes for months. It's the legend on these islands that the only way to remove the smell from your clothes is to bury the offending article in the ground for six months!'

The only bird that may be challenging the fulmar in terms of numbers in Shetland is its much smaller relative the storm petrel. But because people rarely see them no-one knows exactly how many there are. Tiny black sea birds, no larger than swallows, the storm petrels only come in to land during the darkest part of the night, and then only in the summer to lay their eggs and change places with their mates on the nest every two or three days. The rest of their lives they spend at sea, well out of sight of land.

This kaleidoscope of bird life changes dramatically as winter approaches. Northerly gales of incredible ferocity howl down on the treeless islands, and huge seas lash the cliffs. Many of the birds take evasive action, and large groups migrate.

Most of the auks – the puffins, guillemots

and razorbills – stay away from land in the winter, and disperse widely over the sea. In autumn the gannets begin their journey to travel the coast of West Africa; kittiwakes, the real 'seagulls', the most marine of all the gulls, fly right across the Atlantic – young birds are capable of this journey in their first winter; Arctic terns, the jet-setters of the seabirds, leave in search of the sun. 'They travel with the sun. They live in a perpetual summer because they can move right down to the South African coast, almost into Antarctica for the summer, making their way back again here for spring. In Shetland it's not the swallow that brings the summer it's the tern. The terns arrive back on almost the same day each year, having travelled several thousand miles. One misty morning in May you will hear the first Arctic tern, and know that summer has come again.'

Room vacated by the sun-lovers is much sought after by the hardy birds which have been enjoying the rich food bonanza of the Arctic summer. Birds like the huge glaucous gulls come winging down to Shetland for the winter. The smallest member of the auk family, the little auk, which is no larger than a starling, descends on the islands in winter often to claim the title of most numerous bird.

Whooper swans arrive around the end of September, landing, as Tulloch has often observed, on exactly the same spot as the previous year. 'They come back year after year to the little freshwater lochs, and they

Storm petrels may rival fulmar numbers, but the birds only land at night and are hard to count.

Kittiwakes, the most marine of all the gulls, often fly right across the Atlantic.

stay with us until one of two things happens: either they graze themselves out of food in the loch, or the lochs freeze up.' The whooper swan's 'fidelity' to its mating partner fills Bobby Tulloch with a special sense of wonder. 'One of a pair of whooper swans was killed when it flew into a power cable. That was 22 years ago. Afterwards its mate never left Shetland, it spent its summers on a little offshore island and its winters on a croft. Last winter it was killed by a car and by that time it had become so familiar with its surroundings that it could be hand fed by the people in the croft.'

Winter and summer, the impressive list of Shetland birds is augmented by visitors from other parts of the northern hemisphere. 'This is the delight of Shetland for the bird watcher. You really never know what you might find when you go outside in the autumn and the spring. We have birds that breed in the Mediterranean, winter in Africa but somehow overshoot and land in Shetland in the spring. It's a drastic mistake for the bird to make but a great joy for the bird watcher looking for rarities. We've had red-footed falcons and sardinian warblers turn up here!'

Subject of man's most controversial cull – the grey seal pup.

Tulloch himself does not promote this exotic ornithology. He loves birds and would like everyone to share his more simple enthusiasm. 'I think one of the reasons why an interest in wildlife is growing is that it has such a wide appeal, for so many different people, from the pleasure from those who just feed the sparrows on their windowsills, to the person with an intense interest in everything that grows, swims and flies. I think I would count myself in the latter category.

'I am involved in seeing how the whole natural scene keeps going, how it interweaves and how one kind of animal can live with, against or in spite of, another. I number man among the animals because, to a great or lesser extent, we are part of the natural scene – although we do our best at times to make it pretty unnatural. The more remote the place in which you live, the more you are part of the natural scene. It is in our own interests, therefore, to look after the other things that are sharing with us what can sometimes be a pretty harsh environment.'

All his life Tulloch has been aware of this latent empathy between man and nature. A respect for the natural world based on the sharing of a harsh environment is the legacy of the Shetlander. Ever since there have been people on the islands, fishing has been the primary source of food; Bobby Tulloch still follows that tradition, seeing no conservation contradiction in taking enough fish for his needs. On the contrary, he thinks Shetland fishermen, at least until recently, were naturally conservationist. 'They knew that if they took all the fish they would die as well. There is absolutely no justification for over-fishing, it's madness. At the moment some species of fish are being over-fished by the thousands of tons – to their detriment. There is something wrong with the system that allows this to go on.'

He has similar pragmatic reservations about the controversial annual culling of grey seals. 'The idea that we're going to cull the seals because it's "good" for them – because they're destroying their own habitat – I find difficult to accept. I have a sneaky feeling that there's more to it, because if the seals were left alone they would sort things out for themselves. Culling to reduce the population because it interferes with what we do [fishing, for instance] I find very hard to justify. I find it less hard to justify shooting seals if you need something the seals can provide; like the oil and the flesh we required in the old days.

'It's all a question of balance. I don't disagree with the seal cull on the principle that seals are lovely cuddly animals, although it's difficult not to get sentimental about something that gives you pleasure and joy, like the flight of a bird, the speed of a fish and the grace of an otter.'

Bobby Tulloch spends as much time at sea, on patrol in his boat, as on land.

Tulloch was born in a croft with a view of the open North Sea. In winter huge breakers from the force ten gales would come thundering into the bay in front of the little stone house. Here he learnt his place. Incomes were very low, jobs were scarce, people lived off the land and particularly the sea and small boys became naturalists whether they wanted to or not. In Tulloch's case sea birds were a passion from an age earlier than he can actually remember. 'My mother tells of an incident when I was about three years old. I was a constant source of worry: if they took their eyes off me for a moment I shot off towards the sea. I seemed to have an affinity for it. One day they lost me, but after a bit of searching found me dragging home a very dead puffin. They promptly took this away from me, sent me indoors and made me promise to stay there. Later on that night when we had been tucked up in bed in the attic, my mother came to check up on us and noticed a horrible smell.

She turned down the blanket and there was the dead puffin, like a teddy bear. Whether that started my interest in birds I don't know, but it went on from there.'

He went to school on the island, walking a mile and a half each way through a landscape that teemed with bird life. There was a road for a part of the way, but it was seldom used. More interesting was a stream that ran alongside the road and on the way home, young Bobby would fish trout for his tea. 'Sometimes we got permission to come back round the cliffs, although that had to be arranged in advance because it took a long time. But I also had a route which allowed me to get from the low shore, where I was allowed to play, to the high cliffs. It was a very tortuous route but it kept me hidden from the house and parents. It was often used.'

After leaving school, Tulloch had a variety of jobs ranging from apprentice baker to fisherman, even starting a small business 'which didn't work too well because I always wanted to get finished and go bird watching.' Then in the early 1960s tourism started to get underway in Shetland. 'People were beginning to travel around to the north to look at birds. Unfortunately, as well as watchers there were egg collectors and we had one or two nasty incidents in Shetland. At that time the Royal Society for the Protection of Birds did not have an office further north than Edinburgh. I was offered a part-time post to keep an eye on the islands, which I very gladly accepted. In fact, it was a boyhood dream come true and gradually it built up to the full-time job it now is.

'When my hobby of bird and wildlife watching became a profession I felt vaguely guilty, because in the normal Shetland context it wasn't a real job. It wasn't considered practical enough among crofters and fishermen, and I was a bit embarrassed by it. But one night at a dance – and dances in Shetland go on all night – I found two fishermen in an argument. Far from discussing a fishing matter, they were arguing about points of identification between the great grey shrike and the lesser grey shrike: they appealed to me as referee!'

Since those early days, Tulloch has recognized that his job with the RSPB has considerable point: the threat to this remarkable bird haven increases daily. He spends a good part of his time, as do most Shetlanders, on the beaches. But Tulloch is looking for dead birds as opposed to firewood and beaches now depress him. 'It seems like all the world's plastic ends up on the beaches of Shetland. You don't need to go to night school to learn a foreign language – just walk these beaches reading the bottles!'

Bobby Tulloch would be prepared to live with the plastic detritus however, if he could just rid his beaches of the real menace – oil. Shetland has become the terminal for much of the North sea's newly discovered crude oil resources. This bonanza for Britain is so important to a country teetering on the brink of industrial decadence, that nothing has a higher priority – least of all the natural beauty of the Shetland Isles with their wild mammals and their teeming bird life.

Shetland can no longer claim to be unspoilt. Tulloch is seriously concerned that it could also cease to be a safe haven and that would be a catastrophy for the birds of the northern hemisphere. 'We used to think that we were well clear of the problems that had become apparent in the seas of the world which everyone was treating as the sewers of mankind. Then oil was discovered out there – 60

The shag is one of the many birds now threatened by Shetland's new role as an oil terminal.

miles to the east of Shetland – and we've now got a big oil terminal and huge tankers up to 300,000 tons coming in and out of Sullom Voe every day, with all the associated problems that this presents.

'We face the possibility of catastrophic damage being done to what is really one of the better places in Britain for sea birds. And I wish I could believe that the chances of this happening were remote, but it happens in too many places in the world for us to believe we can get away with it for ever. We have already had one oil spill, and it highlighted how even quite a small spill could do considerable damage where you have such a dense sea bird population.

'Only a few weeks after the terminal at Sullom Voe opened, a tanker accidentally struck one of the mooring dolphins, split a hole in its side and spilled about 1,100 tons of fuel oil – not the crude oil from the North sea, but her own fuel oil – into the harbour. A boom collapsed and about 600 tons escaped from the harbour, distributing itself all round the islands in Yell Sound. Within days we were walking the beaches gathering up corpses of birds; corpses of otters even.

'During that spell we found 4,500 dead birds. And not only our own birds. There were 147 great northern divers from Iceland or northern Canada, long-tailed ducks from Iceland and most of the wintering population of black guillemot in that particular area. It demonstrated that we have an international responsibility. Now, nearly two years later, the black guillemot population is only just beginning to show signs of increasing again.'

Tulloch's only consolation was that this spill showed up the deficiencies of the present system of oil transportation. But as well as spillages, Shetland now suffers the effects of deliberate oil pollution, the dumping of dirty ballast. Not all tanker skippers commit this illegal act, but there are still what Tulloch describes as 'floating rust boxes carrying oil on the seas of the world'. Tulloch concedes that the oil men are conscious of the threat and have invested large sums of money in safety precautions. There is a Shetland Oil Terminal Environmental Advisory Group on which Tulloch sits as the representative of the Shetland Bird Club. There is a regular air patrol aimed at deterring tankers from dumping oily ballast: one ship has already been refused entry to the port for commiting this offence.

But given these 'reasonable precautions', what happens when an unforeseen accident occurs? Already there have been two in Shetland. Once a tanker drifted for four days with its engine out of action; on the second occasion, the ship was leaving Sullom Voe when its engine failed. 'Fortunately the tugs got to her in time. These are the things no one can legislate against. As long as international oil politics are what they are, this is going to be hanging over us.'

Tulloch is very concerned that the committee and so-called safety procedures will promote acceptance of a certain level of pollution. Shetland now takes part in the RSPB's National Beachbird Survey. About 100 beaches are examined on one Sunday every month. 'It didn't seem necessary before. For years and years in Shetland I can't ever remember having seen an oiled bird. Now it goes up and down. We peaked after the oil spill in Sullom Voe, now it's back down again to what would seem to be a level we will have to accept. Here I think there's a danger because it's now normal to find big chunks of oil stuck to pieces of plastic on our beaches. Ten years ago that would have been talked about, but we now *expect* to find it. You have to watch where you sit down on the beach here. That's not normal and we don't want it to be normal.

'But I have to be realistic. I run a car and a boat and the oil has to be found for me and millions of others. It's too bad that our wildlife has to bear the brunt of it, but we have to face the fact that this is now the situation in Shetland.'

Huge oil tankers cruise near the Shetland coast, while many birds find food along a shoreline now subject to oil pollution.

As pragmatism is characteristic of Shetlanders, Bobby Tulloch will probably come to terms with normal oil production. But he will never come to terms with his real nightmare – the possibility of a very large oil spill. 'It could be disastrous! The entrance to Sullom Voe is not very far away from the huge National Nature Reserve of Herma Ness where perhaps a million birds are concentrated in the summer. Most of these birds rest in great rafts, thousands of them drifting on the sea below the cliffs. You can imagine what would happen if there was a big oil spill or a tanker collision near that. In terms of wildlife it would make the *Torrey Canyon* disaster seem almost mild in comparison because there were no tremendous concentrations of birds involved.

'Certain species would be disastrously at risk. Puffins, guillemots, and razorbills depend entirely on the surface sea waters for their food and living space. Even birds like red-throated divers, which nest on the little lochs, would be affected because they do all their fishing on the sea, and they winter on the sea. And if the oil were to get onto the beaches as it did the last time, then a much wider range of birds would be hit. In Shetland, particularly in the winter, when the ground is covered in snow and there are storms at sea, a great many birds will take advantage of the area between the high and low water marks which never freezes. It's food supply is always available to birds, including sparrows, starlings and wintering snow buntings. And it's not just the birds: otters feed in the seas here,

and there aren't that many left in Britain. If they have to wade through a patch of oil, when they groom they will be poisoned. In the little spill at least 20 otters died.

'I earnestly hope and pray that a disaster of this magnitude is remote, but the record of what has been happening around the world in the way of big tanker accidents does not give me any reason to be complacent about the situation. And it gets worse when you consider the state of some of the tankers using Sullom Voe. Several have been completely banned by the port authorities; they were in such a pathetic state of general rotten-ness, with untrained crews, no fire-fighting equipment – just rusty old heaps.'

It is desperately ironic that this most committed and knowledgeable of naturalists, whose whole life demonstrates that the best relationship with nature is the one in which you grow up, should be facing such a terrible dilemma.

Of the great number of threatened habitats throughout the world there can be none so tragically juxtaposed as Tulloch, his birds and the oil. It is really not a question of *if* the disaster will happen, but *when*. And, being realistic, while the oil fields are more important to us than the bird habitats, no-one can prevent that nightmare from turning into a reality.

Reference

LINKLATER, E., *Orkney and Shetland*, Hale, 1965.
TULLOCH, R. and HUNTER, R., *A Guide to Shetland Birds*, the Shetland Press, 1970.

For the gannets that feed in large numbers in the surface waters of the North Sea, the possibility of an oil spill is perhaps the greatest threat to their survival.

LEONARD WILLIAMS

Making Sense of Monkeys

The Monkey Sanctuary, Looe, Cornwall, England

Visitors to the west of England, driving toward the seaside resorts of south Devon and Cornwall are likely to observe roadside signs urging them to pay a call at 'the monkey sanctuary'. Those curious enough to follow the signs will eventually find themselves negotiating a narrow and winding lane a few miles from the fishing port of Looe. At the bottom of that lane lies a wooded estate and a country house overlooking the sea; and on the gate collecting the entrance money they are likely to find June Williams.

June is the wife of Leonard Williams, and his partner in an extraordinary experiment in living – or rather in 'inter-living' – because since 1964 the Williams' and their co-workers have been flourishing side-by-side with a colony of woolly monkeys. The sanctuary is not a zoo and visitors will meet only one species of animal, the lagothrix or woolly monkey, from the forests of South America, a species that has a poor survival record in zoos. 'Some years ago we were the first people ever to breed them', June claims 'but now it's like a natural colony with all the different stages of family life that you would see in a wild group'.

In the sanctuary the monkeys do not live in cages as they do in zoos but have a colony territory: five heated indoor houses, large grassed enclosures and a tree area. From the gate monkeys can be seen swinging across a rope bridge from a wooden tower to reach a grove of tall beech trees. Visitors are told that they will have the opportunity to attend what June calls 'a monkey meeting', when they will be able to see the monkeys at very close range, and even mix with them.

Leonard Williams, musician, philosopher and primatologist, with one of his monkey friends.

On a wooded estate in the west of England, Leonard Williams and co-workers established the first colony of woolly monkeys ever to breed successfully and survive outside the South American rainforests.

The big house in the sanctuary has been adapted to the needs of the humans and their monkey neighbours. Although the monkeys occupy part of the house they also have their own territory, the boundaries of which are clearly demarcated by a number of doors and hatches, with access to the outdoor area and to the trees by a complex system of runways and a rope bridge.

The man who created this extraordinary institution is sometimes to be found mingling with the visitors or holding court on the terrace, and those who are fortunate enough to talk to him are not likely to forget the experience. Despite poor health, Leonard Williams remains a vigorous polemicist with controversial views on almost everything. Although his son John Williams, the guitarist, is better known to the general public, Leonard enjoys a formidable reputation among musicians, behavioural scientists, moral philosophers and many unsuspecting persons who have felt the sharp side of his tongue.

Leonard Williams was born in 1910 and spent his younger days in Australia. As a distinguished jazz and classical musician he worked for many years for the Australian Broadcasting Commission. Later, in London, he founded the Spanish Guitar Centre where hundreds of musicians – including his son John – had cause to be grateful for his inspirational guidance and acute critical mind.

How he came to be living and working with woolly monkeys is something that he talks about as if it is a surprise even to him. 'I think that in my case there was a turning point in my life somewhere around the age of 50. I had reached the point when running the school [of guitar music] had become very much a matter of repetition. I was getting tired of living in the city and at the back of my mind was the experience I had in Australia when I worked at the Victoria Park Zoo: it was the first time in my life that I'd done some really hard physical work. I was keeper of the hippopotamus! I don't think we had a clear objective in view. June and I had often talked about how nice it would be to live a little closer to nature, and somewhere at the back of my mind, because of the zoo experience, was the thought that one day we could have a little animal park and, as far as we were concerned, this would have to be with monkeys. Many people have asked me whether we're monkey cranks and I always deny it: we're *woolly* monkey cranks!' Leonard claims that the final decision to set up the sanctuary was taken one

Leonard and wife June: self-confessed woolly monkey cranks!

day at Regent's Park zoo, in London. 'I'd only known June for a few months at the time [June is his second wife], and when she saw her first woolly monkey she fell in love with him right away; at that point we decided to have a monkey sanctuary in the country.'

Leonard and June were joined in their enterprise by June's sister Lorna and her husband John Tucker, and later by a former colleague from guitar-school days, Susan Rickard, and her husband Simon. With two recent arrivals, Daniel Mayer from Mexico and Kathy Day from San Francisco, they live together as what Leonard calls 'an extended family'. He detests the word 'commune' (which has been used by others to describe their group) because of its connotations. As he points out, it is almost impossible to use the word without 'hippy' or 'drop-out' coming into mind, 'so we never have anything to do with the word at all'. The fact remains that at the monkey sanctuary today there is a community of eight adults, four children, a number of North American marmots, a parrot and a couple of donkeys living alongside a colony of woolly monkeys that has grown steadily over the years and now numbers 19.

Most of the visitors to the monkey sanctuary are families with a high proportion of young children but there is no question of talking down to the audience. Members of Leonard's 'family' take it in turns to give simple but detailed explanations of the work of the sanctuary, and try particularly to destroy some of the popular myths about monkeys. Simon Rickard begins his talk by explaining that 20 years ago it was the experts' opinion that woolly monkeys could not be kept in captivity. 'At London Zoo, over a 10-year period, there were 40 woolly monkeys that had been taken as babies from the wild and whose average lifespan was only two years and nine months, which meant that no monkey was surviving even into adolescence.' At that time no woolly monkey had ever been born in a zoo anywhere in the world. 'Here, in contrast, we've been very successful with the monkeys. The eldest member of our colony is 24 years old, she is very fit and active, has had seven children of her own and now has five grandchildren, and she may have more babies.'

The use of words like 'children' and 'grandchildren' to refer to monkeys may offend the purist, but using human family terms is the easiest and clearest way to explain the structure of the colony to a predominantly young audience. The remarkable success that Leonard Williams and his co-workers have had in breeding woolly monkeys has no great

At the age of 21 years, Jessy is the grand old lady of the colony.

At the 'monkey meeting' organized by Leonard and his co-workers, cordial relations depend on humans remembering their 'monkey manners', or risking the consequences!

mystery about it. As Simon Rickard explains 'it's common sense. All we've done here is to provide the monkeys with the conditions they need for a natural and healthy life.' During these meetings some of the monkeys are brought out to mingle with the public, and it is clear that what a woolly monkey needs in order to live that 'natural and healthy life' is a rather familiar formula. In June Williams' words: 'A monkey needs company and privacy, involvement and escape, mates and playmates, space and containment, warmth and comfort as well as air, grass, trees and water.' A better definition of basic *human* needs is hard to imagine.

Given these conditions, the monkey sanctuary is proving that the image of the violent, vicious monkey is a myth. In 20 years there has never been any serious violence between the individuals in the colony and no member of the public has ever been injured by them. During the monkey meetings, the visitors are urged to treat the animals with great respect and they are warned that failure to observe good monkey manners can result in disciplinary action. Simon Rickard's instructions are clear: 'They are not like pets and they are certainly not trained to perform. Treat them as very primitive but very sensitive human beings. Above all don't hold them, pick them up or try to cuddle them, because that is unfriendly. On the other hand, if you offer them a hand or make a comfortable lap for them to sit on or, even better, groom them as they groom each other, you'll find it easy to make friends with them.'

Simon tells a cautionary tale about a man who failed to listen to advice and treated one of their monkeys, Jessy, very rudely. 'First he gave her a rough pat, which she didn't like, then he tweaked her baby's ear, which frightened the baby, and then he tried to pick Jessy up. She couldn't believe his impertinence so she jumped onto his lap, gripped him by the ears and shook him, just as she would have shaken one of her errant children. Of course, the man wasn't hurt but his pride was shattered. He was first told off by a monkey and then laughed at by everyone watching.' The audience always enjoys this story and also the tale of the playful monkey who grabbed a visitor's wig and vanished to the top of a tall tree with it. But judging from their response they also absorb the more serious message of the monkey sanctuary.

Talking to a number of families after their visit we found that they all compared it favourably with other zoos they have experienced. They remarked that under conventional zoo conditions monkeys often appear vicious, dirty and lethargic and they were delighted to discover that the contrary was true; woolly monkeys at the sanctuary never have ticks or fleas, they have a rather pleasant body odour, are gentle and beautiful animals.

When the last visitor has departed and the keepers have assembled in the incredibly jumbled kitchen, the connecting door between monkey and human territory is opened. The most frequent visitor is Jessy. She comes only when she wants to and she stays only as long as she finds it worth her while. She tours the room touching and smelling and paying particular attention to anything new or in an unfamiliar place. Finally she settles near the electric fire, opening her legs and luxuriating in the warmth in a most unladylike manner. When Leonard speaks to her she responds with the high-pitched greeting sound, which he has transcribed as 'eeolk'. He has made a detailed study of woolly monkey vocal communication and speaks the language fluently. When he returns her greeting she abandons the fire, crosses the room and climbs onto his knee. Leonard continues the conversation with an 'ogh-ogh' – another greeting sound but more intimate. Jessy responds by snuggling close to his chest, whereupon Leonard brings his head close to her ear and makes what he calls the 'tuff-tuff' sound. Soon monkey and human are contentedly tuff-tuffing at each other – a way of saying in woolly monkey language, 'I like you very much and I'm happy to be here'.

Monkeys are individuals and have personalities very much their own!

In the Williams' colony, monkeys are free to roam, or just to sit and meditate.

Communication among lagothrix monkeys is just one area of their behaviour that has been thoroughly studied only since the Looe colony has been established. The paucity of information on the animals of the rain forest is discussed in another chapter (page 137), but with monkeys the problem of behavioural studies is aggravated by the fact that they spend all of their time 150 feet up in the dense jungle canopy. Until Leonard Williams began his work almost nothing was known about their social life, and what he has discovered has led him to certain conclusions that go far beyond conventional ethology. Konrad Lorenz, who has long been in communication with Leonard Williams, has described his observations as 'simply superb'. Such warm praise must be gratifying coming as it does from the man who is often called the father of ethology. What is more surprising is to find that the works of Leonard Williams are discussed and reviewed at length not just by natural scientists but by philosophers and political scientists.

To put it far too simply, Williams' observations of the social life of woolly monkeys have led him to some very controversial conclusions about the human condition. In his book *Challenge to Survival* he announces the aim of his work in uncompromising terms: 'to disclose the moral dynamic of the nature of man, and to produce the evidence for a moral imperative that operates throughout the whole of evolution and world history'. He, like Konrad Lorenz, is deeply concerned by what he sees as man's loss of contact with his instincts. But he goes farther and argues that those inherited characteristics are fundamentally altruistic and moral. He claims that 15 years of interliving with monkeys have led him, above all, to rethink his view on the role of aggression. 'It must be realized that there are two kinds of aggression – one natural and healthy and the other pathological – and that aggression in itself both exists and is healthy. Life indeed, whether in a monkey or human society, would be impossible without it. But if that natural power of innate aggression, with which we are born, is corrupted by a sick environment, then it will break loose in various forms of pathological violence.'

Although he is the dominant male, Danny is frequently involved in playfights with other males, even pretending submission (below).

Leonard is fond of quoting one outstanding example of healthy, creative aggression in his monkey colony, which happened when the monkeys were first given access to the grove of beech trees. Leonard and his co-workers had imagined that the colony would be excited at the prospect of enjoying this natural habitat and would instantly migrate there en masse. But such was not the case. Instead, only the dominant male, Jojo, crossed the rope bridge that linked the enclosure with the trees, and Jojo went alone. He behaved with great caution, pausing frequently and never indulg-

Above and Right: Django (a young adult male) invites 8-month-old Ben to take a ride. All adults act as uncles and aunts to *all* juveniles, whether related or not.

ing in his normally exuberant swinging, leaping and rapid climbing. It soon became clear that he was testing the safety of the new territory: every branch was inspected to make sure it would bear a monkey's weight. Dead wood and suspect branches were broken off and dropped to the ground. He tested different routes up and down until he had thoroughly explored and learned each tree. So carefully was the task carried out that it took two whole days before he was satisfied. Until then any other monkey attempting to approach the tree was warned off and chased away.

At the time Jessy was a young and capricious adolescent and she alone managed to give Jojo the slip and reach the trees. According to Leonard 'she cascaded through the branches with gay abandon, fortunately without mishap, though a lot of dead wood cracked under her weight'. Jojo's reaction was rapid and unusually violent. Not only did he chase Jessy out of the trees but he employed the cough-bark, a sound that Leonard transcribes as 'offwharf', a fierce alarm signal very rarely heard at the sanctuary and which he believes is used in the wild to warn the monkeys of an approaching predator. But on this occasion Jessy got more than a warning. When Jojo caught up with her he hurled her to the ground and gave her a thorough shaking. Only on the following day was the whole troupe permitted to enter the tree area without incurring Jojo's wrath. In Leonard Williams' view this apparently violent treatment of Jessy can hardly be termed aggression. 'He was simply drawing upon his instinctual power to aggress in the interest of group stability and survival.'

In her studies of chimpanzees, Jane Goodall has remarked that males never fight over the females, only over their place in the hierarchy. Leonard Williams has found that

All primates have boundless curiosity, and to live happily in captivity must never be allowed to get bored.

woolly monkeys are equally tolerant. Although the dominant male will mate with the female during the optimum breeding period, he is quite unpossessive if she couples with other males when she is not on heat. Moreover, all male and female adults act as 'universal uncles and aunts' to all the infant members of the colony, quite regardless of their parentage. All this, and many other aspects of their behaviour, Leonard sees as 'instinctual social morality' in which aggression – protective and disciplinary aggression – plays an essential part. Only by maintaining a delicate balance between aggression and appeasement is the social structure of the colony preserved. The obverse is also true. If the social structure is disrupted or denied (as it is in most zoos), monkeys develop traits of pathological violence or passivity.

Here, perhaps, is one of many lessons that man can learn from the monkey. Leonard believes that much of the functionless, senseless violence, vandalism and, indeed, apathy found among humans occurs because, like zoo animals, we are 'caged' in totally unnatural urban communities. He suggests that we too, as social apes, do not respond well when we are confined in pairs in semidetached cells in the suburbs. Hence his determination to live as a community with his extended family; by so doing he can maximize the conditions in which natural morality can operate.

Unfortunately, according to Leonard, most people embrace the conscious or subconscious belief that nature is somehow immoral or amoral and that 'our own instincts are representative of the darker side of the human condition, a kind of primitive backwater, the least human and, of course, the least moral from the parson's point of view'. If we look at the whole history of human destruction, greed and cruelty there seems to be very little evidence for the claim that human beings are the only moral animals. On the contrary, Leonard believes that there is a natural morality inherent in the higher animals that man is in danger of losing. 'We cannot of course return to the trees and live like the monkeys, nor can they point the way to our future, assuming that we have a future. Nevertheless, highly socialized animals like monkeys and apes can teach us a great deal about parental care, mutual aid and social responsibility.' Leonard is particularly abusive about the views of behavioural psychologists who reject the idea of in-built social instincts and maintain that we are conditioned into moral responsibility by education or punishment. He speaks darkly of 'those who would prefer to see us as bundles of conditioned reflexes waiting, like Pavlov's dogs, for someone to ring the bell'.

When the weather is cold, the runs and heated indoor quarters are welcomed.

Living side-by-side with monkeys has provided numerous lessons for the humans, not least in child care.

To support his fiercely held views, Leonard is able to call on Darwin for support. The following quotation may come as a surprise to those who associate the author of *The Origin of Species* with 'the survival of the fittest'. But according to Leonard, by the 'fittest' Darwin did not mean the survival of the most corrupt. As Darwin said:

> 'There is no reason why man should not have inherited from a remote period in the past a degree of instinctive love and sympathy for his fellows. Man is a social animal and it is almost certain that he would inherit a tendency to be faithful to his comrade, for these are qualities common to most social animals. As love, sympathy and self-command become strengthened he might then declare – in the words of Kant – "I am the judge of my own conduct, and I will not in my own person violate the dignity of humanity".'

Leonard Williams would probably argue that the most important spin-off from his study of woolly monkeys is contained in his book, *Challenge to Survival*, because he is convinced that our very survival depends on strengthening our natural morality.

Many of Leonard's aims are profoundly and widely misunderstood and to show by what yawning distances minds can fail to meet, he loves to show his friends a couple of letters he received from innocents who fondly believed they were writing to just another zoo-keeper. His replies speak for themselves.

Rowland Ward Limited, Crawley Road, London N22
Taxidermists to sportsmen of the world

Dear Mr Williams,
From time to time we receive enquiries from clients who wish to purchase animal skins of all types, usually made up as rugs from our stocks.

I am writing to enquire whether you can offer us skins of such animals as tiger, leopard, polar bear, etc., when any of your animals die. If you can arrange to supply us, I shall be glad to hear from you in due course.

Yours faithfully,
A. A. Best
(Director).

Dear Mr Best,
In reply to your letter, I regret to say that we have very little to offer you in the way of animal skins. Most of the big game hunters who are shot in our sanctuary are collected by their relatives, who prefer to make their own funeral arrangements.

Recently, however, a distinguished colonel – famous for his sportsmanship as a great killer of African wildlife – was captured by us alive. His skin is in poor condition, but his mane would make an excellent wig for a chimpanzee friend of mine who is going bald. If you can supply me with the wig, I will be pleased to let you have the colonel free of charge.

Yours sincerely,
Leonard Williams.

The Royal Navy, H.M.S. Lynx BFPO Ships

Dear Mr Williams,
H.M.S. LYNX is a 'Cat' class frigate, due to Commission on 11th October, 1976 for service at home and in the Far East.

The Secretary of the Federation of Zoological Gardens has suggested I write to you in an effort to obtain a stuffed mounted Northern Lynx as a trophy for the ship. Any assistance you could give us would be

appreciated and would result in publicity for the ship and the donor at our Commissioning Ceremony.

If you are able to offer any assistance in obtaining such a trophy, I would be grateful if you would contact me.

Yours faithfully,
D. M. Ling, Lieutenant,
The Royal Navy.

Dear Lieutenant Ling,
In reply to your letter of 16th August, there must be some mistake. I am an animal conservationist, not a taxidermist.

However, my good friend Chief Sitting Lynx of the Iroquois Indians in North America has a very good mounted and scalped specimen of a lieutenant of The Royal Navy, which I'm sure he would donate if you mentioned my name.

Please convey my regards to the Secretary of the Zoological Federation and tell him I look forward to seeing him stuffed at the Commissioning Ceremony on October 11th.

Yours sincerely,
Leonard Williams.

Certainly there are many people who regard Leonard Williams as an eccentric and even a disagreeable customer. He has upset a good many establishment figures by his open disdain, not just for most zoos and some current scientific thinking, but for most mainstream politics, economics, sociology and religion.

But even those who disagree profoundly with his conclusions concede that his work with monkeys has been of immense value to conservation. Thanks to Leonard and June and friends, there is now a detailed body of knowledge on how to create conditions where woolly monkeys will live and breed happily in captivity. For the hundreds of thousands of visitors who have visited the monkey sanctuary over the years there has been, at worst, an entertaining couple of hours, at best, the stimulus of encountering a man described, by Alex Comfort, as 'a great original'.

Kate and Joe, the two youngest members of Leonard Williams' extended family, frequently visit their monkey friends; but the human family and the monkey community each have their own territory and the monkeys are never regarded as pets.

Reference

WILLIAMS, L., *Challenge to Survival*, Allison and Busby, 1978.
WILLIAMS, L., *The Dancing Chimpanzee. A Study of the Origins of Primitive Music*, Allison and Busby, 1980.

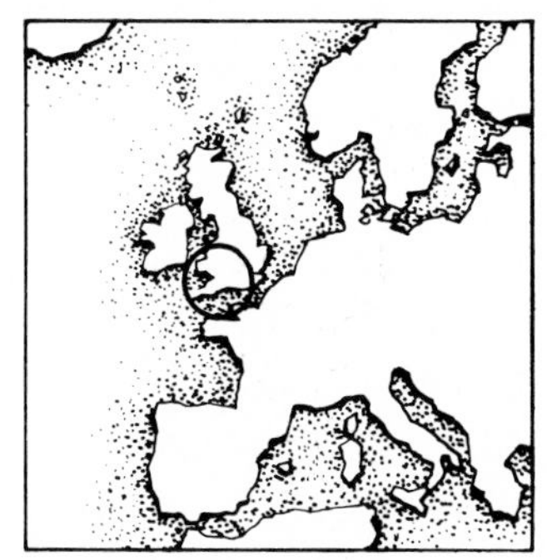

ADRIAN SLACK

Playing God with the Plants of Prey

Frome, Somerset, England

Summer in Somerset; a warm, lazy caricature of an English August afternoon with the bumble bees lolloping about the rose bushes, a fine mist of gnats over the sticky leafed azaleas and a wasp hovering long-legged over the strawberry jam.

Adrian Slack, looking like a cross between James Robertson Justice and Patrick Moore is ensconced on his terrace. He has been talking since mid-morning, pausing only to refill his wine glass. Slack is totally in his element. He is an outrageous anglophile and the day is completely to his liking. The wine is French, he has an attentive audience and the table is loaded down with the strange objects of his obsession – carnivorous plants.

But the closer you examine the scene the more bizarre it becomes. Days like this do not happen in England very often, it is as if Adrian has conjured it from romantic memory. Eccentrics like Slack, complete with cream linen suit and Panama hat, come from that same romantic past. And there he sits, a committed vegetarian all his life, completely seduced by plants that eat flesh.

Therein lies Adrian Slack's considerable drawing power. He is an anachronism – a throwback to the great era of English naturalists when cultured men with small goatee beards patrolled the English countryside with picnic hampers and butterfly nets.

It is a carefully constructed and quite deliberate image. He lives and works in a remote part of south-west England where it is perfectly possible to believe that time has stood still since Edwardian days. Indeed,

Adrian Slack: vegetarian obsessed by flesh-eating plants.

Giants of the carnivorous plant kingdom, the yellow trumpet pitchers, growing naturally in a place that echoes the conditions in which life began, the warm wet Okeefenoke swamp in southern Georgia, USA.

Ancient Marston Mill in Somerset where Slack manages to grow tropical pitcher plants.

apart from a few details, such as the shape of the tractors and the TV aerials, it might well have done.

Until we stepped through his time door, Slack had deliberately avoided serious movement outside his Somerset frame of reference. His home is there; his greenhouses are a few miles away; his social life is shared with friends of like persuasions. He runs his own supper club in a pub several centuries old and the only partially political group he has ever joined is one that periodically refights the English civil war.

He is, however, a founder member of the British Carnivorous Plant Society, an organization he helped to create, and he did, in 1980, produce the best and most authorative textbook on carnivorous plants yet to be published in English. That much of him is completely real even if it is a little unreal that he should have managed to pick up all that expertise on his own: he is both 'untaught' and 'unqualified' in any formal sense. He describes himself as a 'plantsman'.

He would like to be seen as a kind of mid-Edwardian Gilbert White, on the botanical side of the family. In fact, he is rather better than that. He does not need to be a caricature: his love and knowledge of carnivorous plants make him an original. If he epitomizes anything it is the committed amateur naturalist whose enthusiasm eventually ranks him, or her, with the most qualified professional natural history scientist.

The study of natural history was a tradition spawned in Britain, and later taken up with alacrity by underemployed English priests (Gilbert White was the vicar of Selbourne) and genteel women who were also rather short of things to do. There are few words in the language quite so evocatively English as 'naturalist'.

We should, therefore, have been less surprised when the offer of a trip to the New World, there to visit a carnivorous plant Eden called the Okeefenoke swamp, was not at first greeted with the enthusiasm we felt it deserved. We did not know that Slack had never been outside England and to leave its shores would, like a chameleon in space, involve him dropping some of his highly protective colouring. But the idea eventually grew on him (had not Burton entered Mecca and Darwin the Galapagos?). The man is the most industrious talker we have ever met, but for a moment the 'American difference' took his breath away. He is the most gregarious individual we came across, hunting receptive listeners like bees seek honey, but for half a day American friendliness rendered him positively shy. Fortunately, the sights, smells and plant life of the extraordinary Okeefenoke swamp, otherwise known as the 'Land of the Trembling Earth', cleared the American log jam and the words began to flow again.

We went in at dawn, poling through channels of dark peat water between tangled screens of swamp plants. Slash pine, known locally as turpentine trees, stick up out of the water, their tops hung with ghostly wraps of spanish moss. As the sun rose, huge cobwebs spun from fine strands of copper glistened across the 'hoorah bushes'. Once the Okeefenoke was a centre for moonshine whisky production in the South and when the 'revenuers' made their raids, the stills would be hidden in the swamp bushes. Relieved moonshiners would mutter 'hoorah' from the thickets as the revenuers passed by – or so the locals would have us believe.

The Okeefenoke is technically as close to Eden as you can get on our planet, one of the few remaining echoes of the kind of conditions in which life on earth was developed. There is plenty of water; it is warm to hot all the year round; there is perfect cover.

Small wonder that the Okeefenoke National Wildlife Refuge boasts at least 200 species of woody plants, most in great abundance. Many of these are exotic and present a spectacular display almost all the year round.

Feeding on these rich growths, and on each other, are a mass of animals. At the last count there were some 18,000 alligators in the swamp, numerous species of snakes, black bear, a host of small mammals and literally millions of insects.

Slack in his element, a warm English day, a glass of wine, and a crowd of carnivorous plants.

And there is the Okeefenoke's unique collection of indigenous plants that have one foot, or root, in both the animal and the plant kingdom – the plants of prey.

By this time, with the dawn yellowing all around us, Adrian was almost incoherent with excitement. Rather than having to search for the pinguiculas, sarracenias and droseras, they were everywhere. It started as we approached the park gates in the first light. Slack spotted something from the car and we were commanded to stop. We found him with damp knees beside the road. There in a line alongside the tarmac were bladderworts that absorb animal protein through bladder traps on their roots, miniature trumpets that attract insects with nectar then lure them to their death, and very tiny sundews that threaten the insects of the microuniverse with a sticky end. Only the prospect of giant versions of these vegetable predators growing wild on the floating islands in the Okeefenoke, eventually lured Adrian back to the car.

In the swamp itself, it was Adrian who spotted the first clump of hooded pitchers with a shout of alarm: 'Good God! That shouldn't be there at all.' The object of his amazement was some three feet tall, sallow green, and rising to a reddish-speckled hood, like a rearing snake. There were several of them and they looked ominous.

There is, in fact, no good reason whatsoever for humans to be even marginally afraid of carnivorous plants, but there is no denying that they have 'monster' appeal, simply

Okeefenoke dawn. Slash pine hung with Spanish moss loom out of the still swamp water; a perfect habitat for insects and their vegetable predators, the carnivorous plants.

because they do live on meat. Their food is one of the smaller forms of animal protein – insects – but it is flesh and blood none the less. This 'flesh-eating' ability also gives the carnivorous plants a special status in the human mind. Normally we see plants as part of another world; they are not even remotely related to us, and their lifestyles and life-processes are alien to animals. They do not have hands, lungs, mouths or stomachs and they cannot move about.

Except for the carnivorous plants. Venus's fly traps have clawed extensions with spines that 'grab' their prey much as we would catch a butterfly. Yellow, hooded and parrot pitchers have openings very like mouths leading

Bladderworts that capture insects in tiny traps are grown by Slack in Somerset.

In the Okeefenoke swamp, bladderworts abound, floating about in search of prey.

down into chambers where their meals are digested in a not un-human way. Butterworts – looking like small green starfish – can move: their leaves curl inwards over their prey. And in the Okeefenoke some of the bladderworts can move long distances: they have evolved a raft on which they drift about, trailing their traps in the water to catch aquatic insects.

It is almost as if this is a branch of the plant kingdom that is creeping up on us, even though that impression is botanical nonsense. The hooded pitcher plant (*Sarracenia major*) is one of the worst offenders. Its snake-like hood, protecting a honey-coated throat and a maze trap, does look predatory. Even though Adrian Slack has been interested in carnivorous plants for more than 20 years he still admits to a degree of anthropomorphism. 'I do look upon them almost as animals, which of course they are not.' He nodded in the direction of the offending Okeefenoke hooded pitcher. 'Just look at it. It's very hard not to think of it as a sort of meditating monk.'

The existence in nature of plants that are more than inanimate vegetables – 'that actually got into motion and grasped their prey in no uncertain way' – was the factor which launched Slack, as a very young boy, into a hobby destined to become his life's work. He had been given an illustrated book (it is still on his bookshelf) in which there was a

The most obvious of the carnivorous plants, the trigger-action Venus's fly trap.

water colour of a plant with round leaves, red tentacles and glistening dew drops on the tentacles. 'I wanted to get that plant', Slack remembers. 'They said it grew in bogs and I went all over Herefordshire looking in any place that was moist. I went up the hills and even into the woods, but alas, the mission was hopeless. Year after year went past and still I looked for that plant. I didn't actually find it until I was 18.' The plant was the common round-leaved sundew, and by this time Slack was laying the foundations of a career as a landscape gardener.

'Having got one, of course, I wanted to grow it. The first time I tried I failed to grow one, but I managed the second and that was the beginning of my collection, although, of course, I didn't realize it at the time.'

He badly wanted some of the more exotic American carnivorous plants and eventually acquired them – albeit somewhat illegally. 'A kind person came over on the Queen Mary and brought me, illicitly in a sponge bag, a whole lot of plants; in particular my first Venus's fly traps.' The Venus's fly trap (*Dionaea muscipula*), although only a few inches high, is certainly the most impressive performer of the carnivorous plants. It grows 'claws' more than an inch long. Adrian Slack studied his illegal immigrant with a sense of wonder that he still feels today. But then a

Contrary to legend no carnivorous plant, even a clump of Venus's fly traps, is in any way dangerous, other than to insects.

Venus's fly trap does have extraordinary abilities.

It attracts flies and other insects by its coloration – the insides of the traps are a pretty coral pink deepening to red – and by producing nectar. Inside each trap are fine hair-like triggers which, when activated, cause the trap to close on its prey. But the mechanism is complex. First, a trigger will not activate the plant unless it is touched twice within about 20 seconds, or two separate hairs are touched once within the same period! In Slack's opinion this double trigger is an energy-saver and a selection mechanism. 'If you're an insect and have settled in the open trap and are walking about eating your nectar, you are bound to touch a trigger twice eventually. But the plant doesn't want to be shutting on every bit of wind-blown matter that hits a trigger in passing.'

And double-trigger selection is only stage one of the Venus's fly trap's very picky nature: the traps do not close up tight straight away, rather the spines fringing the two faces of the trap close to form a grill. 'If the triggers have been set off by a very tiny creature of no use to the plant it can, in the next 20 minutes, crawl through the loose grill between the bristles. In this event, after 20 minutes the leaf will start gradually to open again. On the other hand if the insect is substantial – too big to get through the bristle grill – the leaf will start to close in on it, pressing it to death, and will then remain absolutely closed until the creature has been fully digested.'

Adrian Slack, like everyone who has ever seen a Venus's fly trap in action, wondered how a plant which has no nervous system and no brain could operate such elaborate mechanisms. In fact, by careful experiment he has found that Venus's fly traps are selective chemically as well as mechanically.

If a piece of wood or a stone sets off the trap and the fragment is too big to fall through the bristle grill, the plant will close down on it. But when it fails to get what Slack terms a 'nitrogenous signal' from the crushed object, the plant does not produce digestive acids and enzymes and will, after about 24 hours, open again. Slack proved that the Venus's fly trap actually conducts a chemical analysis of trapped items. 'If you put some meat extract on the stone, you find that all the digestive processes begin.'

But even though he has been gazing at

Venus's fly traps with wonder and through microscopes for two decades, he still has no idea what causes the reactive processes of this extraordinary plant. He suspects the impulse messages which cause both a build up and a loss of turgidity in cells along the spines of the trap lobe are electrical.

Now hopelessly seduced by the exotic contents of the Queen Mary sponge bag, Slack became a serious collector and the landscape gardening assumed second place. To understand how carnivorous plants had evolved, he needed a great many species and his collection soon became very unwieldy. 'I went up to Cumberland for three years, and the Bishop of Penrith had a lot of mouldering greenhouses which he allowed me to do up and take over. Eventually there were hundreds of plants. Some of them had to go outside; I was surprised to find that some survived this, but in the end I was throwing them out by the handful. We had huge funeral pyres in the garden: it was pathetic.

'At a dinner party I was telling a friend of mine how depressed I was after burning plants all day and he said "Why don't you sell them?" ' Thus began one of the most successful cottage industries in contemporary Britain and a perfect solution for Adrian Slack. The Somerset greenhouses and the unwieldy collection were now financed by mail order sales to an ever-expanding public. It also meant that Slack could play with his plants, hybridize them and breed selectively.

But to do that he had to learn all there was to know about the many different types of insect-catching plants, and how they worked. There are 14 different genera of carnivorous plants, best classified according to their method of trapping prey, which is either active or, more commonly, passive.

A fine stand of yellow trumpet pitchers.

The passive trappers can further be divided into three groups according to the type of trap employed. For example, 'pitfall' traps are characteristic of all five genera of pitcher plants – sun pitchers (Heliamphora), trumpet pitchers (Sarracenia), cobra lilies (Darlingtonia), tropical pitcher plants (Nepenthes) and the West Australian pitcher plant (Cephalotus). The 'lobster pot' variety, which involves a one-way portal, is employed by one of the sarracenias, the parrot pitcher (*Sarracenia psittacina*) and a unique water plant,

Genlisia, of the bladderwort/butterwort family. The third variety of passive trap operates like a sticky flypaper and is used by two genera, the rainbow plant (Byblis) and the Portuguese sundew (Drosophyllum).

Active traps are employed by seven genera and can be divided into three types. The first has added a twist – literally – to the sticky flytrap: the sundews (Drosera) of which there are more than 90 species, have evolved minute tentacles which curl round their prey. Similarly, the butterworts (Pinguicula) can curl the tips of their sticky leaves round a trapped insect. The remaining two mechanisms are the 'steel traps' and the 'mouse traps'. The Venus's fly trap is the best example of the former; it shares the method with another genus, the waterwheel plant (Aldrovanda), which has a tiny trap of only 2 mm that operates underwater. Mouse traps are used by the bladderworts (Utricularia), of which there are more than 250 species, and the prettily named pink petticoats (Polypompholyx), of which there are only two species, both in Australia. Prey are sucked into these traps through a door levered open by bristles.

As far as the evolution of carnivorous plants is concerned, the most primitive representatives, the sun pitchers, were first discovered in 1839 on the top of mountains in the deep south of Venezuela – 'an almost unclimbable range that became the model for Conan Doyle's *Lost World*.' These fleshy green plants were little more than rolled leaves that secreted nectar and collected rainwater in which insects searching for the nectar conveniently drowned. From the rolled leaf, the pitcher plants went on to develop better nectar glands, tiny vegetable umbrellas to protect the nectar from rain, and even drainage systems to prevent the pitcher filling with water and becoming top heavy.

Over the millions of years of their evolution, the pitchers became experts in the business of bribery. Nectar was carefully spread down the plant along nectar trails, the supply increasing as the insect reached the pitcher opening. Some pitchers evolved and mixed a soporific drug with their nectar. 'They get positively tight.' Slack relates, 'the insects become very unsure on their feet and zoom down the pitcher where it gets very slippery and they can't get out. But I'm convinced they die in a very happy state!' Yet another pitcher deliberately collected rainwater in which to drown insect prey. But as a great number of insects can escape an accidental ducking, this particular pitcher (*Sarracenia purpurea*) has evolved a wetting agent which it secretes into the water making it impossible for insects to lift off.

Combining all three devious devices are pitchers which collect lethal water, make drugs, and have evolved the equivalent of a maze to trap prey: semi-transparent 'windows' in the plant wall act as false exits to lure the wandering insect ever deeper to its doom.

Among the most deceptive, and attractive, is the great cobra lily (*Darlingtonia californica*). This pitcher's most singular feature is a fishtail-shaped 'tongue' that is covered with nectar and is very attractive to insects. 'They lick their way on towards the richer supplies of nectar until they reach a little mouthpiece. Above this, inside in a kind of a dome, there are the false windows. Insects, after partaking of all they require, will head for those windows, thinking they are exits; they go back and forwards trying to find their way out until eventually they reach the slippery bit at the bottom – and plummet down.'

But the most sophisticated and complex of

The most predatory looking pitcher, the cobra lily. Nectar trails and false windows in the stem of the plant lure insects to a slippery place from which they plummet to their doom.

all pitcher traps must surely be the parrot pitcher (*Sarracenia psittacina*), so called for its distinctive parrot-like 'bill'. There is only one small entrance (it is impossible not to think of it as a mouth) into a parrot pitcher, and, as with others of the family, they lure their prey into the mouth using a trail of nectar bribes. The parrot pitcher has also evolved two convenient grooves in the direction of its mouth which some insects may follow out of convenience (although nectar is the main lure). Once inside a parrot pitcher the insect is in real trouble. 'There is then a little tube which goes right into the inside. They go down that tube, again towards windows. They try all these little windows all the way down the pitcher to no avail, and without realizing it they pass through long thin hairs, criss-crossed.

'The hairs can be easily pushed apart going down but because of the way they are crossed there is absolutely no way back for the insect. They are caught in just the same way as a lobster is caught in a lobster pot. It is quite astounding!'

Although the complex mechanism involved is indeed remarkable, Slack is well aware that the pitchers are among the most simple of the carnivorous plants: no matter how ingenious their traps, they are still *passive*. However, several varieties of carnivorous plants take a much more active part in capturing their prey. Among these are a group of almost ethereal beauty if they are studied under sufficient magnification – the sundews.

Sundews are like Christmas-tree decorations in miniature. In place of leaves, their hair-like branches carry microscopic droplets of sappy 'dew'; tiny jewels of liquid that attract and refract the light. They are lethal traps. The simple ones act like flypapers. 'The fly lands and is immediately covered with dew. It struggles and will fall, accumulating more and more of this sticky material until all its breathing tubes are completely clogged and it expires. Tiny glands that are so small you can only see them under a microscope then come into play and start secreting the digestive fluid round the fly.' More advanced sundews have made a great leap from the flypaper stage, evolving tentacles to hold the struggling insect in the dew trap. Cape sundews from South Africa and

forkleafed sundews from Australia are examples of these; they also fold their leaves over the trapped prey to increase the speed of digestion. Slack acknowledges that were the sundews any bigger they might well be a little frightening – 'like those awful things one reads about in American comics'.

The plants that share the 'active flypaper' system with the sundews – the butterworts – would never frighten anyone. In fact of all the carnivorous plants they are the most innocent in appearance, with bright-coloured flowers, and 'leaf-shaped' leaves. But as the great Charles Darwin demonstrated, the pretty butterworts are also sophisticated insect killers. Insects are caught in a sticky liquid secreted from the leaf-surface, and as the insect struggles, glands pump out more mucilage which eventually suffocates the prey. A second set of glands then comes into play, secreting a liquid which slowly digests the prey and absorbs the resulting nutrient fluid. To assist with digestion, by bringing a greater number of these glands to bear, the butterwort's leaves curl inwards to cover the dying insect completely.

And, finally, to the bladderworts, which set traps that Adrian Slack extols as: 'by far the most ingenious of all'. Bladderworts, by any definition, are very odd plants. They have no roots and it is often impossible to tell the difference between stems and leaves. They can have flowers so plain as to be inconspicuous, others so beautiful as to be called 'fairy aprons'. One thing they all have in common are their ingenious bladder traps; tiny, no bigger than a pin head, but masterpieces of natural mechanics, and lethal.

Seemingly as innocent and beautiful as a Christmas decoration, sundews are lethal flypapers.

Sophisticated simplicity: bladderworts trap minute insects in tiny underwater bladders.

The bladders are attached to the plant at one end by a narrow stalk, while the other end is taken up with a very special door, complete with a 'lock and key' mechanism, plus machinery for impelling an insect into the trap! All the bladder traps operate in water. In the Okeefenoke swamp, bladderworts float around on little rafts, fishing with their bladders. Terrestrial species use their bladders in the wet subsoil.

Prey are urged towards the trap door by antennae that form a kind of tunnel. The trap door is, in fact, a valve, the top part fixed like a hinge, the lower hanging free. The whole structure is kept watertight by glands which secrete a mucilaginous sealant, and a membrane which locks the seal shut. Four bristle levers point in the direction of the approaching prey.

The concave sides of the bladder maintain a slight vacuum inside the trap and when an insect hits one of the bristle levers the whole mechanism is so finely integrated that the faintest push (the smallest utricularia can catch single-celled protozoans) opens the door, resulting in prey and water being flushed into the bladder. The easing of the vacuum causes the door to flip back into place. The water is then sucked out of the bladder, restoring the vacuum and within half an hour to two hours the trap is ready to work again. Once the food is inside the bladder, internal glands secrete digestive enzymes and acids.

In the process of assembling the material for this book we have come across a great many things in support of the homily 'isn't nature wonderful!' but none to really compete with the extraordinary traps of the innocuous-looking bladderworts.

Small wonder that Adrian Slack has been hypnotized by these carnivorous plants and their brothers and sisters, since first he saw their mechanisms under a microscope.

Nowadays, he spends most of his time 'playing God' in his greenhouses by creating hybrids to enhance the appearance and reputation of carnivorous plants. He takes every opportunity he can get to talk about them with evangelical intensity. 'One cannot perhaps improve upon nature but you can get a plant which grows better in cultivation, is easier for the amateur to grow, is more colourful and has a finer form. I've been doing this for over 20

Playing God in his greenhouses, Slack is able to breed and hybridize many carnivorous plants.

years now, and I'm getting some rather good plants. It's a fascinating thing which I cannot, I shall never, tire of.'

It also has a somewhat more serious purpose. Adrian Slack, who has lived in the English countryside all his life, is well aware that plants are less well protected than animals. 'Apparently no day goes by without some plant becoming extinct somewhere in the world. A lot of these carnivorous plants are threatened.

'But I do believe some sense will prevail in the end and in the meantime species are being kept alive "in captivity". Some of the American pitchers are practically extinct in nature, but fortunately some of my friends have kept them going and now are replanting them back into the Appalachian mountains.'

Adrian Slack, fortunately, is a comparatively young man and is hardly likely to be tempted away from his almost idyllic existence in Somerset. It may seem bizarre that flesh-eating plants of the tropics have found a haven, with a vegetarian, in a greenhouse in south-west England, but it hardly matters if the existence of the species is at stake.

They could not be in better hands.

Reference

SCHNELL, D. E., *Carnivorous Plants of the United States and Canada*, Blaire, 1976.

SLACK, Adrian, *Carnivorous Plants*, Ebury, 1980.

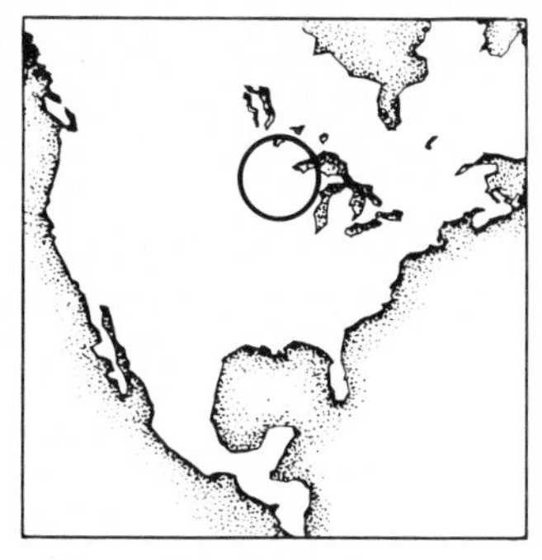

GARY DUKE AND PAT REDIG

In-Flight Repairs

Raptor Rehabilitation and Research Service, University of Minnesota, St Paul, Minnesota, USA

Of all the wild animals on this planet, birds are the ones we take most for granted. We do so because birds have made an extraordinarily successful choice of habitat in a world dominated by man. More simply, if a fox were to leap over the wall into your urban, or for that matter suburban, garden, it would be an event. If a leopard were to do so, you would have your name in the paper. The fact that many species of birds do so all the time is pleasant, but hardly worthy of comment, even though most of them are just as wild as the fox.

In the depths of our most appalling concrete jungles it is still possible to live alongside wild things – birds. Natural history owes a lot to birds for that reason: in many cities birds are all that is left of nature, and a great number of naturalists, amateur and professional, date the beginning of their interest to urban encounters as children with birds.

The bird can live safely all over human territory because, unlike that fox, it can use the sky. In addition birds are generally small and offer little competition to man. When they do, as with flocks of starlings or city pigeons, their immunity sometimes comes to an end.

The bigger the bird, the more subject it is to human interest: an interest which is very often aggressive. As a result, the most threatened birds on earth are also the largest, and if you are both large and aggressive it would seem your days are definitely numbered on the land surfaces of this planet.

In North America this has become so much a reality that a special hospital has been set up to repair human-damaged large birds, the raptors (eagles, hawks, owls and falcons).

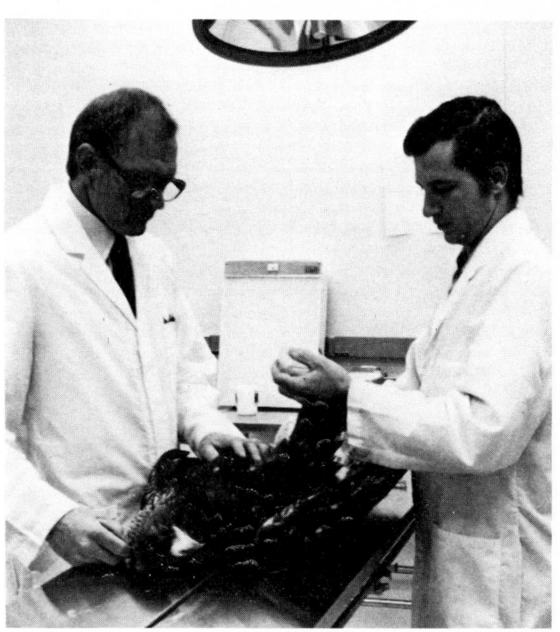

Drs Gary Duke (left) and Pat Redig (right) run a hospital for injured American birds of prey.

The bald eagle, national symbol of America and most protected bird, yet threatened by the shotgun.

This is the University of Minnesota's Raptor Rehabilitation and Research Service, run by Drs Gary Duke and Pat Redig assisted by a research graduate, Dr Eileen Bird. (For the record, every joke possible about Eileen's name has already been tried.)

The bird hospital is no academic research project: damaged creatures arrive there daily by train and car; they are also being flown in by the Fish and Wildlife Service from all over the United States.

On page 121 we take a look at the problems facing Wyoming, a part of North America that is still mostly virgin wilderness, and suggest that if the place is protected now its prolific wildlife may survive. The Minnesota bird hospital, however, casts serious doubts on conservation prospects based on early protection.

All birds in America, with the exception of house sparrows, starlings and pigeons, are protected in some way. The lowest rung on the protection ladder is that you must have a licence to shoot; but you do not have to go too far up the ladder before you meet bird species that no-one may touch – the 'threatened species' – and above that the top echelon of protection, the 'endangered species'.

Throughout the United States, the bald eagle and the peregrine falcon in particular are considered 'endangered'. This in spite of the fact that they are most strictly protected. In fact, the bald eagle is so 'untouchable' that it is illegal even to stare at their nests from closer than 1,000 feet. Drs Duke and Redig have to obtain two permits just to treat the birds that are sent to them: one to cover the rehabilitation; the other to carry out research on a bird that may die.

Behind this protective fence one would assume that such striking birds of prey as the bald eagle and the peregrine falcon would be thriving. Not so. Duke and Redig conclude that 75 per cent of the injuries they treat in the hospital are in some way the results of human interference. Most are caused by collision with man-made elements in the environment, such as cars, power cables and windows, but a large proportion are the result of shooting or trapping.

'It is our personal count of injuries we are treating. We x-ray every bird that comes in here, irrespective of the injury. Now we may be treating a bird that's been hit by a car, but I don't think I'd be off-base if I said that 80 per cent of the [bald] eagles we handle are carrying shotgun pellets.' That means that at least eight out of ten bald eagles, America's loyal symbol and its most endangered and protected birds, get shot at least once in their lifetime.

The case of the bald eagle is a very ominous omen for American wildlife. This beautiful bird is in real danger of becoming a symbol of another kind: a hallmark of man's inhumanity to non-human species.

'It's very unfortunate that of all the birds of prey we handle, the bald eagle seems to be taking the greatest brunt of the shooting', Pat Redig commented, 'but it does tell us something about people: they are pretty callous and not very concerned about their activities in the field or the devastating effect they have on other forms of life.

'I don't believe for a minute that any of the shootings of raptors are accidental. People have known for many, many years that these birds are protected. There is *no* good reason for shooting them. They are just thoughtlessly shooting for the sake of shooting and trying to kill something.'

It was to accommodate this reality, the

realization that protection laws and wishful thinking were not going to protect American birds of prey, that persuaded Dr Gary Duke to convert the University clinic into a fully fledged (sic) bird hospital. Today, with Redig and Eileen Bird helping full time, the hospital has an operating theatre that would put a number of human hospitals to shame, a preparation room, full sterile facilities and a custom-built block for rest, rehabilitation and recreation of repaired raptors.

At first glance the facility seems almost excessively well-endowed. But when you realize that its entire function is to repair the damage done to birds by man – and, to our knowledge, it is the only one of its kind – it seems a very small investment on the part of humankind.

A day in the life of this accident ward is remarkably similar to those familiar television serials featuring Dr Kildare, except that here the patients are birds.

Peregrine falcons were rendered almost extinct in the United States by the use of DDT and DDE pesticides.

09.00: Drs Redig and Bird scrub up for the first operation of the morning schedule. Patient: a red-tailed hawk with a fractured wing, resulting from a blow with a blunt object, or a collision while in flight. It is not that they are accident-prone, but 'in their native environment there are now so many things put up they're not equipped to look out for. Fifty per cent of the cases we receive are accident-type injuries: birds that have been hit by cars while scavenging on road-kills or have flown into things – poles, wires, high tension power lines and windows.

'You have to realize that when a bird is hunting its eyes are focussed on the ground, not on something immediately in front of it. They're not used to looking out for power lines. And in level flight the bald eagle, for instance, is flying at around 50 to 55 miles an hour. If you hit something at that speed you are going to cause severe damage to your body, and indeed they do. Many of the eagles we have here now are amputees – birds that flew into power lines. They didn't just break their wings, they sheered them right off!'

The red-tailed hawl is anaesthetized, its wing plucked in chicken-like fashion, and then carried through into the operating room where a steel pin is inserted through the broken joint. The prognosis is good.

Duke and Redig maintain a fairly consistent 60 to 70 per cent success rate with their orthopaedic operations. However, survival rates also depend on the state of the bird when it arrives. 'Sometimes the wound has been open for a long time and the bone is infected with bacteria; those cases have about a 50–50 chance of surviving.'

10.00 Redig and Bird do their ward rounds. In a series of distinctly aromatic rooms, large and small owls sit on perches and blink their multiple eyelids. Hawks – red-tailed, peregrine and gos – share a rehabilitation 'flying room' and allow human visitors with a surprising lack of aggression.

In some cases the birds have been there a very long time. The red-tailed hawk's fractured wing they have just repaired will take six weeks to heal. The hawk will then have to sit it out in the convalescence room for at least another month. In fact, a bird with a fracture of any kind is seldom released back to the wild in less than five months. During that time the team keep a careful check to see that the bone joint stays free, the bird becomes 'generally reconditioned' and the muscles strengthen. Raptors, they have discovered, are more prone to arthritis than humans.

Next door is the eagles' flying room. Here the doctors pause. 'If we switch off the lights before we go in', Eileen suggests, 'they'll accept our presence if they suddenly find us there'. It is a nerve-wracking experience, entering a room full of eagles in pitch darkness in the knowledge that the door has been well-bolted behind you. But Eileen knows her birds. They stare with what looks like unbridled fury and clench their huge yellow talons, but we all leave unscathed.

Raptor-doctoring can be hazardous, however. Both Eileen and Pat carry talon scars on their hands. The big eagles and hawks can 'fix' right through a heavy leather glove if they decide to. Redig once had to walk up a flight of stairs into the operating room and anaesthetize an eagle to free a set of claws fixed through his hand.

10.30 Duke is called from Wisconsin. A bald eagle has been hit by a car and is presently on the midday flight in a Fish and Wildlife Service container. He alerts the operating theatre and heads for the airport.

11.00 Two small children, accompanied by their father, walk in with a cardboard box containing a robin. Pat Redig puts on his white coat and carries out as careful an examination as he made on the hawk. The robin has a broken wing, damaged when it flew into the house. It is a compound fracture of the elbow, and Pat informs the family that the chance of a repair is not very good. 'But we appreciate you bringing it in. Check back in a few days and we'll tell you what the progress is.' The hospital encourages the public to bring in any bird they find.

12.30 Gary arrives back with a very smelly eagle. Pat and Eileen, apparently immune to the aroma, examine it carefully – and gingerly. She firmly holds the huge talons, while he keeps a careful eye on the beak. 'The trick is to check that someone is holding the talons',

Gary Duke and Pat Redig prepare a damaged raptor for surgery in an operating theatre that would put many human facilities to shame. Seventy-five per cent of the injuries they treat are the result of human interference with birds and their habitats, many due to shooting. None of these shootings is accidental, even though all the birds are protected. The survival rate for orthopaedic operations such as this depends on the bird's original condition, but is usually about 70 per cent.

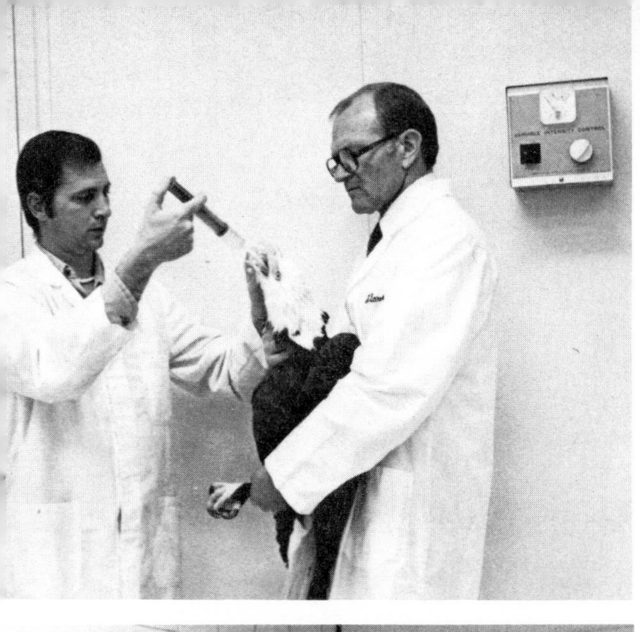

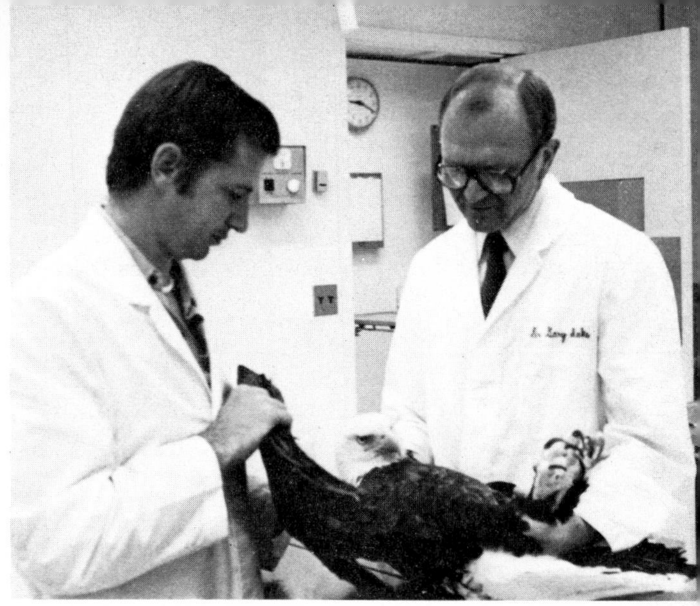

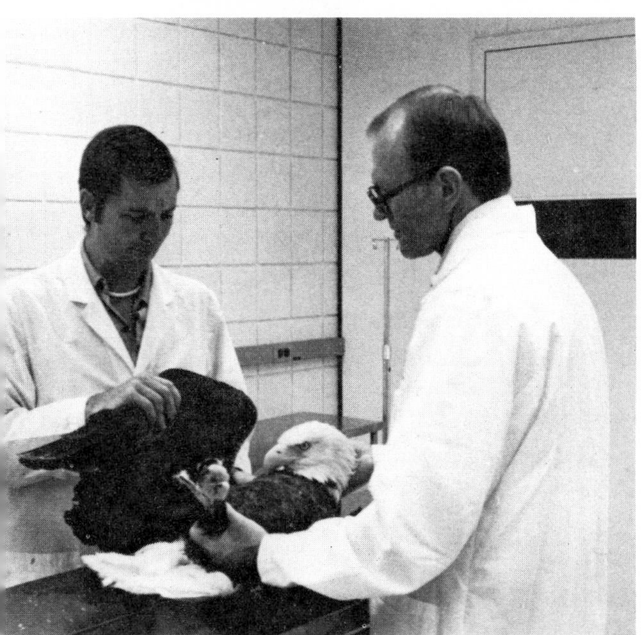

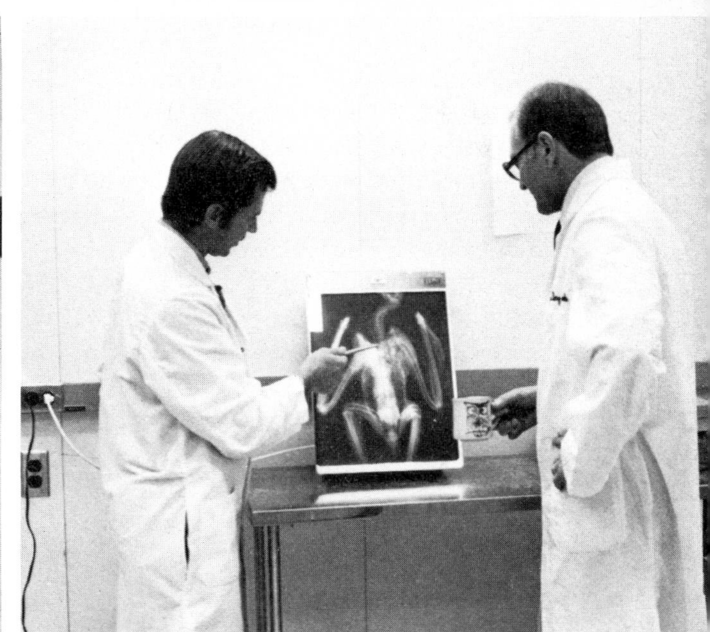

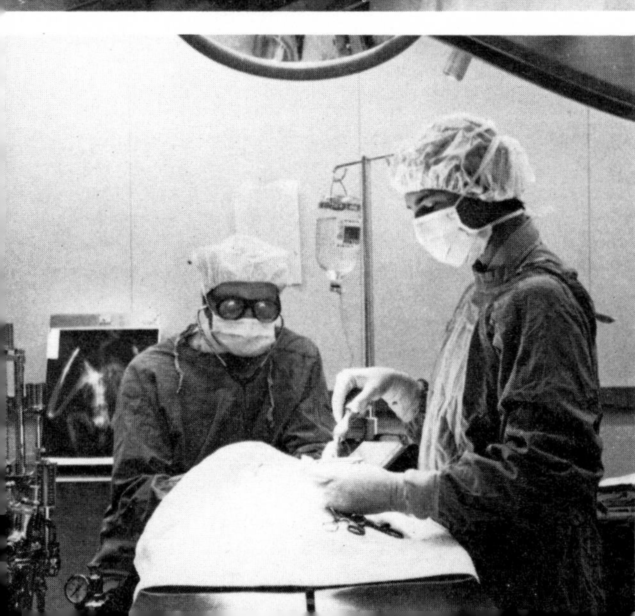

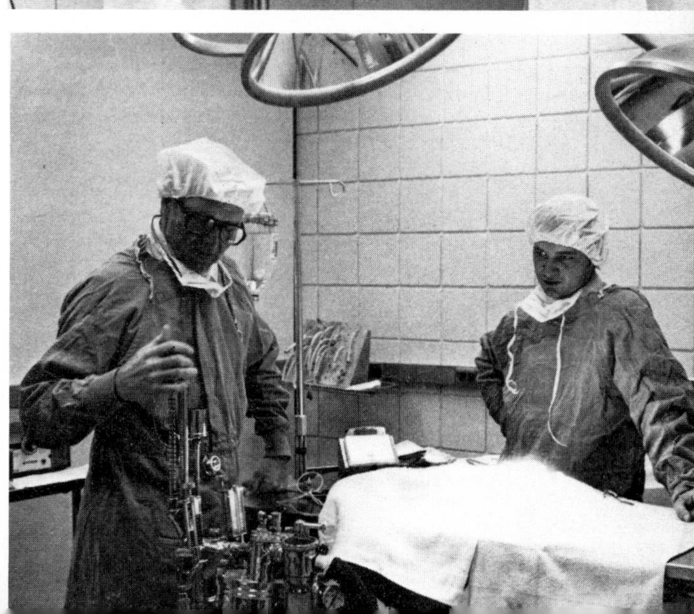

Eileen advised. 'Sometimes you assume the other person is doing it and end up full of holes!'

The eagle looks reasonable. It has a damaged eye and is somewhat thin. It appears not to have a breeding patch on its belly and may therefore be too old to be part of the endangered Wisconsin breeding population.

Breeding birds are the most vital. 'Once we got a female bald eagle from North Carolina. We really went to work on her when we had worked out that we were handling one-sixth of their entire breeding population.'

But all birds, whether young or old, eagle, hawk, robin or owl, get the same careful attention. 'After all we're only guessing that this eagle is not an active member of the breeding population. Perhaps if we get her back into good condition she'll start to breed.'

Stress from accident and the new, confusing surroundings of the hospital is a real problem with birds of prey so no further work is done on the eagle at this time. Blood samples and x-rays are taken, and she will be given fluids and rest. 'We try to minimize the stress for the first 24 to 48 hours. About 15 per cent of the birds that come in die before we have a chance to do anything with them and it seems that if a bird doesn't have the 'live-ability' to survive for 48 hours there's probably little we can do about it.'

13.00 Lunch. Gary and Eileen head off to the University gym for a session of weight-lifting; Pat goes home carrying a plastic bag containing a dead white rat.

14.00 Cleaning: when the students are on holiday, budgets force Pat and Eileen to hose out the hospital themselves. Meanwhile Gary makes some notes on his turkeys. It is important work for him because it contributes a part of the hospital's budget, and since it is of direct benefit to human appetites, it is well funded. But a poster composed for him by his students expresses a rather different view: 'He who would soar with eagles should not work with turkeys'!

15.00 A 'bird mother' arrives with four baby long-eared owls in a box. Gary Duke explains the bird mother service. 'Every so often we get a call from someone who says he has an injured hawk in his backyard but he's afraid to pick it up. We have a number of what are called bird mothers around the city who seem prepared to help even in the middle of the night. They go round and collect the bird, hold onto it until the next morning, and then bring it over to us.'

Bird Mother Barbara's baby owls are another tragic story. 'The two adult birds were shot and we've brought the babies down here to be released where it's a bit safer.' The box is levered open and four clicking, spitting, pecking blobs of angry feather are hauled out for their medicals. One jumps up on top of the box and puffs out its wing feathers to give an impression of increased size, but within seconds, goes distinctly limp. 'Around here they stop doing that after a couple of days when they realize it isn't going to be of any value and we're going to grab them anyway', Gary comments. He grabs and the baby owl fixes its sharp beak firmly round his finger. 'Don't make any mistake about these little guys or any other owls. They make lousy pets. They may look fuzzy and cuddly, but they're not. They're messy and not friendly and don't want to be held.'

The making of pets out of infant birds of prey is now recognized by the Minnesota team as a major problem, and not just because these pet animals can be killed and crippled by misfeeding. The problem lies with the phenomenon of 'imprinting'.

Imprinting is the propensity of birds to make parents of the first thing they see after hatching. Normally it is the parent birds, but it can be humans who have taken baby birds as pets, or worse, eggs to hatch at home. The birds become attached to humans, not just paternally but, in some cases, sexually, once they have reached maturity. Even if a young bird of prey is released, as often happens when the wild nature of the baby begins to show, it will remain attached to humans and may even try to mate with them. Many 'attacks' on humans by owls and other birds of prey are the result of imprinted birds trying to mate. Inevitably such birds are killed, despite the fact that it is humans who are to blame for the attachment.

A similar problem exists with half-tame birds: Barbara and Gary will have to be very careful how they return the young long-eared owls to the wild. 'There's a fairly standard procedure called hacking out – an old falconry term. We put the birds in a large cage in an old farm building, get them accustomed to that spot and feed them there at the same time every day. After they have been there six weeks or so and they're ready to fly you just open the door to the cage or the building, but still keep putting food back at the same time each day. Generally the birds come back and get their regular food, but at the same time they're learning to take care of themselves and getting their food in the normal way. After about a further three weeks they start coming back every other day, then twice a week and finally they don't show up at all. The hope is that they've returned to the wild and are

Owls that take a fancy to poultry often run foul of farmers and end up in the Minnesota bird hospital.

looking after themselves, and, of course, that they're not imprinted on humans.'

Thanks to man, however, it does not always work out as Gary would like. 'During this time the birds are dependent on us for food and are half-tame', he explains. 'Sometimes they go elsewhere for food and get a bad time instead.' (For 'bad time' in this case, read bird-shot.) Barbara explains: 'The bad time comes from people who are trying to protect their game or poultry. Such people sometimes develop a real hatred for predatory birds. That's why we'll release these orphans down here on a game sanctuary or a reserve. We know that where they came from there are several people who shoot owls. The adults were shot: it's very likely these would be too.'

Barbara leaves to look after a screech owl she is in the process of 'hacking out'.

15.30 Pat and Eileen inspect the red-tailed

hawk on whom they operated that morning. To their delight it is standing up and very much alive. When they began operating on birds there were grave difficulties with anaesthesia tolerance: it varied considerably not only between species but also between individuals of the same species.

Their next stop is less happy: the bald eagle from Wisconsin is dead. Pat is not particularly surprised, he had earlier received the developed x-rays and suspected that the bird's encounter with the car had resulted in severe internal injuries.

We worked out that this particular bald eagle had had a very large sum of money spent on it if you counted car trips, the special crate, air flight, medicines, hospital facilities and the time of about seven or eight people, game wardens, airline staff and those at the hospital. But Pat insisted that such time and expense was not for nothing. 'You don't write it off that glibly. We have the carcase of the bird and that will be subjected to a complete postmortem. We'll take tissue samples that will give us information about pesticide intake, and we took some blood. What we learn will help us with the next bird.'

A major study conducted by Pat Redig on dead raptors points yet another accusing finger at the dangers of excessive hunting in the United States. He discovered that large numbers of eagles and hawks were suffering from lead poisoning: these birds are scavengers, and Redig realized from his examination of the stomach contents of dead birds that in certain areas they were scavenging wildfowl that had been shot, but not recovered, by hunters. 'Winged' ducks that lazy hunters had not bothered to recover, or birds used for target practice, were being tidied up by the eagles going about their normal food hunting. But these wildfowl carcases were contaminated with lead shot: a slow poison for eagles. Redig's study has been taken seriously by some ammunition manufacturers, who have recently started experimenting with stainless steel shot.

Birds, even baby owls, soon learn that aggression displays do not attract much attention in the rehabilitation centre.

16.00 Playtime for raptors. Pat, Eileen and Gary load boxed eagles, hawks and owls into the back of their station wagon and head for the University football field.

It is alongside a main road which could easily develop the worst accident rate in the state: passing drivers can hardly believe their eyes when Eileen rushes along the touchline, whooping and waving her arms at a hobbling bald eagle, who expresses his resentment with widespread wings and shining steel-like talons. Gary lobs a disdainful horned owl aloft. It can hardly be bothered to flap its wings and sits in the grass glowering at him.

Pat opens a box containing a red-tailed hawk and the opposite happens: it makes a dash for freedom in the direction of the traffic and all three give chase to head it off before it can reach the road.

'This is not quite as ridiculous as it may seem', Gary puffs, 'except that we're doing it. Usually we leave it to the students, but they're on vacation'. In fact, all this riotous behaviour is part of the birds' exercise programme. 'When an athlete breaks his leg in the course of a season, he can't take the cast off and immediately return to the field. The same is true of these birds. When we release them, they have to be fit enough to catch their prey – they must be in peak athletic condition.'

Pat Redig's goshawk, Olivia, foster-mothering red-tailed hawk chicks from the hospital.

A great horned owl faces the moment of truth. Is he fit enough to go free?

It seems, however, that the team is getting more exercise than the birds!

17.00 Home time, and we are invited to Gary's to 'meet Matthew', and Pat's to see 'Olivia feeding the babies'.

These are not bizarre extensions of their families. But it is inevitable that two such committed naturalists should have taken their work home. Fortunately, they both have bird-loving wives.

Matthew is Gary and Mary-Anne Duke's 'house-owl' – not their pet owl, mind. Matthew is firmly imprinted on Mary-Anne. He came to her as a baby that had been taken out of the nest and badly raised by a couple of young boys. He had seven fractures and rickets and Gary was convinced he would die. There was no way he could get round-the-clock medical care at the University in those days, so Gary brought him home and Mary-Anne nursed him back to health, becoming, through imprinting, Matthew's mother.

'He had two wings in casts to start with. We got him over that, although his wings were too badly damaged for him ever to fly again. The rickets left him with one usable leg which unfortunately developes sores. The poor little thing was just lying on his tummy in the bottom of his cage for three weeks, severely depressed. But he got through all that and now he seems to be quite content to be a member of the family.

'But before you even think of keeping an owl, face up to the idea of changing the newspaper in his cage every day, vacuuming up a new crop of feathers every morning and cleaning up the mouse intestines from last night's dinner. Plus, you have to have a supply of mice. Our deep freeze is full of them.' Gary insists that the owl is not a pet, and that they would not keep him at home were it not for the fact that he needs so much care and attention. But it was equally obvious they were very fond of him.

Pat Redig also has a damaged baby owl, and, like Matthew, it too has nowhere else to go as a result of being kept in captivity and fed on beef liver. When it came to Pat he discovered that both its legs were fractured and there were five wing fractures. The owl is now the reason Pat takes a rat from the hospital every lunch time.

In cages down his long garden, Redig also keeps hawks. He is a licensed falconer, using his birds for sport and research. 'Olivia' turned out to be his favourite gos hawk. Not flying because she was in moult, Pat had put her to work: she was acting as an excellent mother to two red-tailed hawks from the hospital. Pat was absolutely delighted that Olivia had taken on the job, because it solved all the problems of imprinting with the baby hawks. 'By letting another hawk rear them during the convalescent period, we avoid the problem of human imprinting. Both babies had fractures – we think they were blown out of the nest. One had a wing fracture, the other a broken leg, which meant they would definitely have been imprinted at the hospital and it would have been impossible to release them. Cases like that usually end up in zoos.

'But with Olivia agreeing to look after them, even though she's a gos hawk and they are very nearly as big as she is, we've got two young hawks that in 10 days or so can be easily and happily released to the wild.'

Redig also has a peregrine falcon and this too will be returned to the wild, after experiences at the hands of man which make unpleasant reading. It has taken Redig more than two years to get the bird back to full health and he is openly nervous about where he should stage the release. 'In the northern and mid-western states this is our most endangered bird of prey. Look at her! It's one of the most sleek, elegant creations that's ever flown the face of the earth. Yet we very nearly rendered it extinct with DDT and DDE and those other chlorine pesticides. By and large across the breadth of the country, the peregrine is the most endangered bird and its the species that is the best indicator of impending doom in respect of the pesticides poisoning our whole ecosystem.

And on that ominous note our day in the life of a bird hospital ended, but the most joyous aspect of the work done by this unique team was yet to come. Three birds were due for release the following morning.

Gary and Pat have set up a series of 'safe'

release ranges. Some of these are inside wildlife sanctuaries, but they also have a list of conservation-conscious farmers who are willing to have a recuperating raptor as a temporary guest on their land.

The big high fliers like the hawks and eagles usually lift off and fly miles when released, but owls prefer a stretch of safe cover until they have got their bearings. As two of the birds due for release are owls, the team has chosen one of their safe farms, just a few miles from the hospital. The release site is an open grassy plain with a river nearby and two thickets of young trees. The birds' reintroduction to the wild begins.

Gary's owl flies 300 yards before settling in a bush. He is given another chance and repeats the performance. 'I don't know whether it's laziness or something else' Gary says, 'but he goes back to the ward until I'm sure.' Eileen's owl makes it. A big-horned owl, it shoots off in level flight, weaves round a barn and vanishes in a thick grove of young trees.

Pat Redig's red-tailed hawk is sheer magnificence! It leaps powerfully from his glove, beats the wind with strong wings, catches a thermal at the edge of the field and circles back with the sun burning through its spread feathers, before lifting away from us.

There were no cheers. It was one of those moments when speech catches in the throat and silent wonder takes over. Only when the hawk was 600 feet above us, still climbing the thermal, did Gary Duke break that silence.

'She's been locked up for nine weeks. But get them out on a fine summer's day, and they just soar like that for the pure joy of it. They must do – what other motivating factor could there be?'

The moment of release: Duke and a repaired redtailed hawk fit and ready to return to the wild.

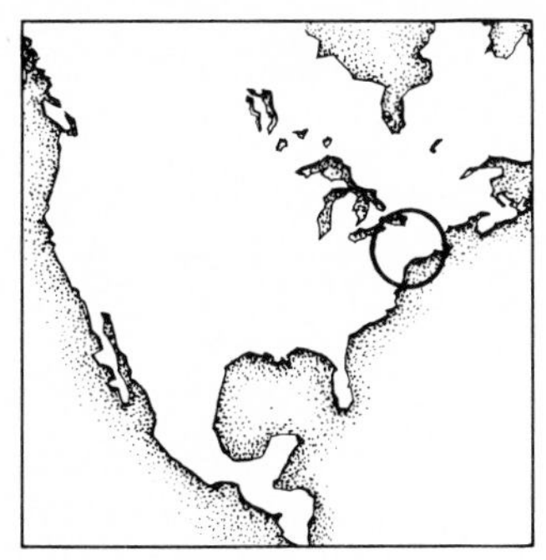

TOM EISNER

An Insect's-eye View

Cornell University, Ithaca, New York, USA

Our view of nature, our likes and dislikes, the things we champion and the things we neglect, is very largely conditioned by the way in which we see.

We have a soft spot for animal features which echo human characteristics and will completely 'humanize' animals given half a chance. Thus dolphins with a genetic upcurve to the mouth are said to be 'smiling', chimpanzees get dressed up in 'baby' clothes and are given 'tea parties', and brown and black bears that can be greedy killers become 'teddies' in human toyland. The latter is particularly bizarre considering 'teddy bears' are named after President Teddy Roosevelt who had no compunction about slaughtering large numbers of grizzlies and North American black bears.

Such anthropomorphizing is probably yet further evidence that the naked ape is not quite the well-advanced, sophisticated beast we like to think we are. The fact that we still prefer animals that look like us, or smile as we do, is almost certainly a hangover from times when as members of the jungle we needed to view all living organisms with suspicion and assess their potential as predator or prey. More simply, we are still frightened of aliens, or animals with non-human features, like slimy toads or for that matter, in reality, dry, horny toads.

Our variable attitude to animals based on perspective applies to size as well as shape and is another ancient and unnecessary fear-echo

Eisner's obsession with the chemical defence mechanisms used by insects has revealed the most sophisticated wargames.

Most cunning of the crafty arthropods, this green lacewing larva (centre) first disguises itself in woolly aphid 'wool' as protection against the guardian ant, before feeding on the aphids.

from our primal past. We still fear elephants and the rhinoceros. Quite understandably, you may suggest; certainly that is the first reaction. But think about it. The only human that has anything to fear from a modern elephant is the indigenous meat hunter with a bow and arrow or a spear, and there are precious few of those about today. In fact, they are probably as threatened a species as the elephant itself. Only a tiny percentage of the world's population will ever get to see an elephant or a rhinoceros in the 'wild' conditions of a game park and these will be sitting in fast buses or Land Rovers with an expert guide at the wheel ready to take off at the first sign of danger.

But show a picture of a charging elephant on television and a frisson of fear will tickle the necks of human viewers sitting safely and comfortably in their own homes. 'Big is dangerous' is the code that comes to us from our primal past.

Small – particularly very small – is variously dangerous, totally unimportant and irrelevant to human existence, and alien. And yet there are more such creatures on earth than any other.

The English biologist, J. B. S. Haldane is alleged to have commented that 'God's purpose in creating the Earth was to accommodate His inordinate fondness for beetles'. In general, humans do not possess that fondness. The most compassionate animal lover will pluck ticks with loathing and swat a mosquito. Flies are dirty the human rubric goes, and may be murdered indiscriminately. Given the vast number of flies in the world, if this were really true, we would surely all have died long ago. Without a moment's pause, a shred of guilt or a second thought we have given ourselves an unlimited licence to kill in the most highly populated game reserve on this planet.

Fortunately, the insect kingdom is so well populated, and many of its inhabitants such successful breeders, that unthinking human predation has had little effect. In any event slaughter is a fairly common occurrence in the insect world, and attack and defence are probably the two most important features of an insect's existence. How good they are at both defines success as a species. When you live in a world that teems with life to the degree where it would seem God's real purpose was experimental biology, you need to be very good at attack and defence just to survive, let alone to be successful.

In Eisner's view the insect world is a great unknown, the counterpart of outer space.

Humans know something about insect attack mechanisms. Spiders, scorpions, ants, horse and deer flies do sometimes make the mistake of sucking human blood or gouging out a scrap of human flesh. They die for it in huge numbers and insects as a whole get a worse reputation than they deserve. But because the insect world is so alien and is largely invisible to the upstanding naked ape, we know next to nothing about insect defence mechanisms.

This perhaps is why the studies and findings of the Cornell biologist, Dr Tom Eisner are so intriguing. Eisner champions a corner of nature that most of us obliterate with thoughtless footsteps. In that inner universe of insects he has found a Pandora's box of fascinating devices, systems and potions.

He found this naturalist's treasure trove by using a different perspective. For Tom Eisner, small *is* beautiful, and it becomes even more beautiful when enlarged under a microscope. Exoskeletons are objects of engineering interest, as are multilensed eyes on green stalks.

Cornell University, in northern New York State, is an Ivy League establishment with one of the largest biology faculties in the world. Eisner is laboratory and lecture-bound for a large part of his year, but believes that the right place for a naturalist is in the field, watching. If you are watching for insects you have to watch in a special way, with an 'insect-eye view'.

Eisner is tall, gangly, a reformed smoker whose nerves have yet to settle, and very talkative with an easy smile. 'We try to imagine what it is like to have to survive as an insect, to find food as an insect, to find a mate, to find a place to lay eggs; to look at things the way an insect might want to look at them. You have to simply look at nature and let your curiosity roam wherever'.

'All we have to do is look under a leaf to discover a whole world that's new.'

Eisner's personal fondness for beetles began during a much-travelled childhood. His father is a German Jew whose search for an alternative to Hitler took him and his young son through Spain and South America. Insects were the ideal 'pet' animal for the infant naturalist; they took up little space and travelled well. Young Tom Eisner also visited places where the insect life was extremely varied and exotic.

Although insects take first place over music in Eisner's life, at Cornell he conducts a university orchestra.

In Uruguay, Eisner started to collect bugs until his room became a 'hazard to the rest of the family'. Also, he discovered another lifelong interest which makes his department at Cornell quite unique. He started music lessons, and today his laboratory contains a piano in regular use, while his home in the woods outside Ithaca is described as 'basically a music room with lounging and cooking space'. The big sitting room actually accommodates two full-size grand pianos and a harpsichord. Every fortnight, Eisner conducts Cornell's 'Brahms' orchestra 'The By-weekly Rehearsal Association of Honorary Musical Scientists'. 'You don't actually have to play an instrument to get into this department,' Eisner's secretary, Peggy (a jazz singer) smiles, 'but it sure helps.'

As a pianist Eisner is no slouch. He was trained by the conductor Fritz Busch. In Eisner's late teens Busch noted that his young student's ear was not quite perfect, an unlucky break for Eisner but a fortunate one for natural history science: because Eisner is not just a good entomologist, he has made a quite unique study of the business of survival in the insect world.

Take the case of the larva of the green lacewing, which Eisner has labelled 'a wolf in sheep's clothing'. This is a good subject for the lay reader to consider because the 'sheep' in this case are minute aphids – the bane of the gardener.

Study one species of aphid under a microscope and you will discover they have white, 'woolly coats' not unlike those of sheep or poodles. Aphids also have another sheep-like quality: they process vegetation in a way that produces a tasty by-product, in an aphid's case 'honey-dew'. Honey-dew is much favoured by ants and they will literally herd and protect a group of aphids in much the same way as a shepherd tends his valuable sheep.

Enter the 'wolf' in the form of the larval stage of the green lacewing – an inexorable predator on woolly aphids and a surprising freak of evolution considering that aphids are protected by ants that are more than a match for green lacewing larvae. To survive in this dangerous evolutionary niche, the larvae needed to come up with something very special in the way of disguise.

Tom Eisner discovered that they had, but even he could not quite believe it when he first saw it. He watched the green lacewing larvae eating aphids and wondered how they were getting away with it. With a sense of wonder approaching total incredulity Eisner realized, on closer inspection, the larva was not at that stage eating the aphid – but plucking its woolly coat. Only when the larva had completely covered itself with aphid wool, and was totally indistinguishable from the aphid flock, did it start to kill and eat. As Eisner's picture (page 86) shows, the larva was ignored by the shepherding ants because of its woolly disguise.

Eisner's interest in the way insects protect themselves may have something to do with the fact that he has always seen himself in the role of a protector of insects and their reputation. It is an uphill task. Even the most committed animal lovers still call insects 'bugs' and 'creepy crawlies', and people who would have very definite reservations about squashing a frog or even killing a bat (neither very popular animals) will swat a fly or stamp on a cockroach without compunction.

Tom Eisner accepts that much of this stems from the fact that insects are so physically different from mammals. But probably more relevant is the fact that most insects are either small or minute. They live out their lives in a Lilliputian world of which we have little or no awareness.

Eisner, a regular visitor to that microuniverse, is completely intrigued by their organic armouries and systems of chemical defence, and keen to publicize these wonders. He is an excellent writer and has become one of America's leading microphotographers.

Sometimes insect systems and their associated chemistry cross over into the world of man: the assassin beetle (*Apiomerus* sp.) provides a good example. Assassin beetles, as the name indicates, are adept killers in the insect microuniverse. But they are also cunning chemists and seem to care a lot about their eggs – an unusual feature among insects.

Tom Eisner noted that this exotic red and yellow beetle had a particular liking for the camphor weed plant, which produces copious bubbles of camphor, a substance we have been using for centuries to keep insects out of our clothes. Close and extended study of assassin beetles on camphor weed plants revealed that they were apparently scraping camphor jelly from the leaves and passing it down their bodies using their several legs. Watching for some days, Eisner eventually noted that this accumulation of camphor jelly was being spread quite deliberately across the abdomen in a thick sticky layer. But

The exotic assassin beetle is a vicious killer and yet a caring mother in a world where maternal affection, or concern for the young, is hard to find.

for what reason or purpose was a mystery. Then the assassin beetle, a female, began to lay eggs, and Eisner noted with delight that each new egg was projected through the camphor jelly, so acquiring a thick coating of a substance that was obnoxious to other insects. The beetle had been collecting an insect-repellent to protect its eggs!

Everywhere Eisner looked he found ever-more extraordinary chemical defence mechanisms being employed by insects in the interests of their own or their offspring's survival. In many cases these mechanisms existed in insects that were well known.

He made a study of Australian termites and quickly established the various defence roles in the termite colony. But did the soldier termites have any weaponry? Eisner constructed a bionic enemy – a sliver of metal of termite size that he could spin in threatening fashion using magnets. As soon as the mechanical 'ant' began to spin in Eisner's laboratory, soldier termites descended on it. Under his microscope Eisner spotted that the soldier termites had a conspicuous pointed nozzle on their heads, and from these nozzles they were seen to be spraying a sticky repellent fluid onto the spinning metal ant! The secretion also acted as an alarm system, and within a matter of seconds there was a ring of soldier termites hosing the mock predator.

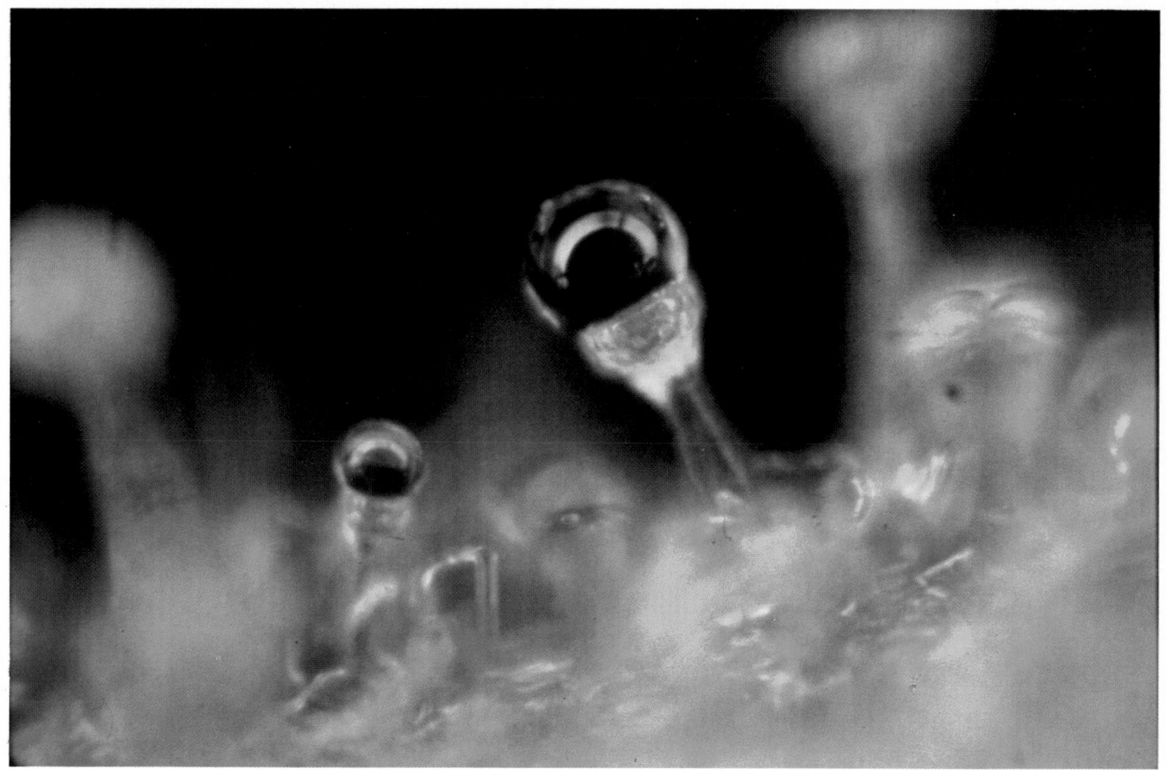

Assassin beetles have discovered that the sap of the camphor weed plant is an efficient insect repellent: they harvest the sap and coat their eggs with it.

Eisner noted that the use of the spray was not automatic. When more than one soldier was in position they would wait before spraying to assess the effect of early shots on the predator. The spray has the consistency of rubber cement and its initial effect on a real ant is to cause irritation. The ant will break off its attack to try to cleanse itself of the sticky spray, but the cleaning process simply has the effect of spreading the glue over the ant's whole body. Moving about in some desperation it becomes coated with grains of soil and other detritus, and is soon completely incapacitated.

Another animal that Eisner discovered could produce a highly effective protective glue was the common slug – an animal which we usually regard as the most vulnerable and 'sluggish' beast in nature. As have many before, Eisner wondered why slugs produced slime. Applying his 'insect-eye' view, he imagined himself as a slug predator and contemplated how this Eisner-insect would go about his attack. The assault, a sharp bite or jab, could come from any direction. The naked slug appeared to be totally unprotected. Eisner prodded a slug with a sharpish implement and found the head of his probe covered with a blob of slime-glue. Merely stroking the animal produced no slime. If Eisner prodded the slug in a different place, another slime blob blunted the probe. It seemed that the slug, if aggressively provoked, could produce slime blobs anywhere on its body. Eisner recognized the effectiveness of this defence mechanism and was soon able to prove and photograph it. He took pictures of ants with their mouthparts gummed-up with slug glue. The slug, like everything else in the micro-universe, was not as innocent and certainly not as defenceless as it might seem. But then it probably would not exist if it was.

This truth emerged very early in Eisner's studies of the insect world. Nowhere else in nature does the dictate 'survival of the fittest' apply so dramatically. Habitats are crowded and ecological niches competed for literally to the death.

Tom Eisner was therefore intrigued to discover that the mechanisms he was studying were essentially defensive, and this raises an interesting question. If an animal, say the slug, is capable of producing a slime that incapacitates a predator why not add a toxin to the slime to kill the predator and avoid future attacks from that source? No one knows that answer, any more than we know why most large animals display more aggression than they practise. But it is somewhat surprising to discover that in the cut-throat competition of the insect world, creatures mainly limit the weapons they develop to the defensive.

Perhaps the classic case of this is the 'star' performer of Tom Eisner's menagerie – the bombadier beetle (*Bracherus* sp). This beetle occurs in a number of countries and varies considerably in size. Those from Kenya are

Even the slug has acquired some chemical weaponry – a glue secreted at will to gum up the mouth parts of predators, such as ants.

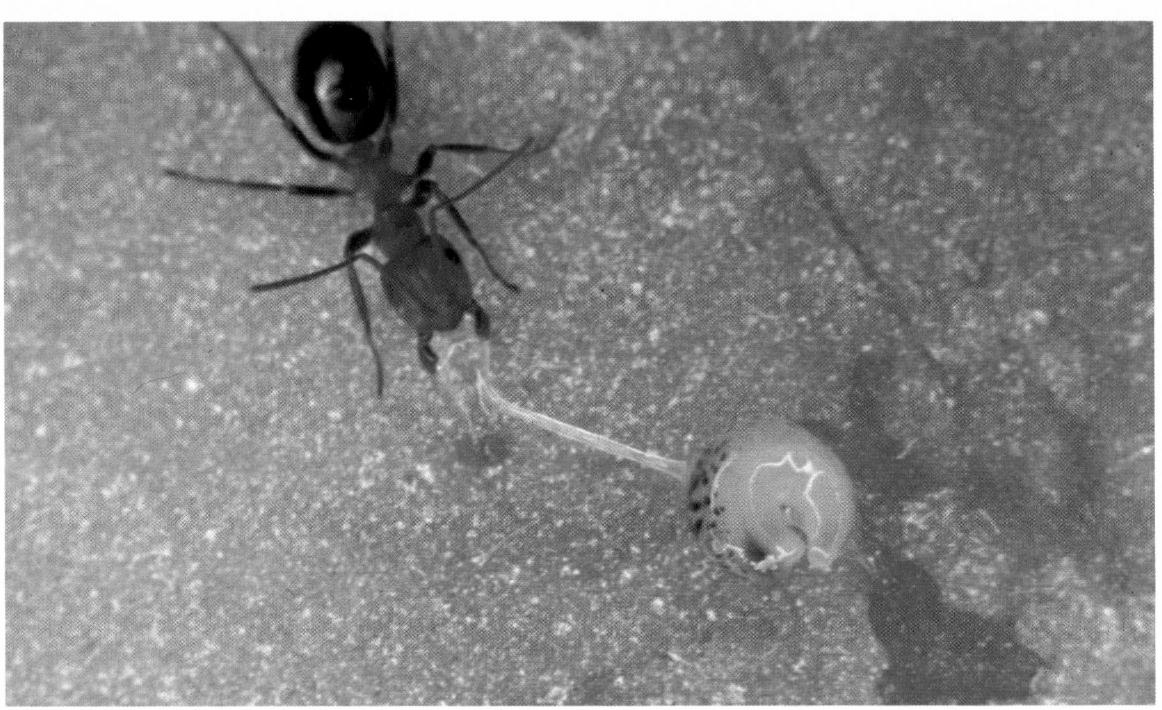

A slug's response to attack depends on pressure: touched firmly it will produce a mass of sticky mucus to ward off the offender.

more than an inch long. Bombadiers from Florida, where Eisner gets most of his specimens, are no longer than a few centimetres. But big and small they are all supreme experts in chemical warfare and weaponry.

Stated at its most simple, bombadiers have evolved a chemical spray to deter attackers. But Eisner and his colleague at Cornell, Dr Dan Aneshansley, have shown that it is anything but simple. The spray contains extremely strong irritants, and the bombadier has evolved a way of heating these substances to boiling point – 100°C! Considering that bombadiers, like all insects, are cold blooded, Dan Aneshansley realized that the beetle must have a very remarkable chemical plant inside its tiny body. Dissection revealed a double chamber not dissimilar to a rocket engine. An inner reservoir held the raw materials of the noxious spray, while the second chamber contained activating agents. The beetle squeezes fluid from the reservoir into the 'reaction chamber' and, as with the rocket engine, an explosive series of reactions takes place, creating heat in the process.

It might be thought that so complex a chemical defence would be enough for the little bombadiers. But to go with the spray it has developed a quite remarkable gun. A rotatable nozzle at the base of the abdomen can direct the spray in any direction: over the back, under the legs, and to the left and to the right to cover the legs. Eisner also established, using high-speed photography, that the spray is delivered in a series of very fast pulses. Why the bombadier should go to this trouble no one yet knows. Tom Eisner suspects that it could be to allow split seconds of cooling in the reaction chamber. Without some cooling process, a bombadier beetle manufacturing boiling spray under heavy attack could easily cook itself.

Eisner had made comprehensive studies of the effect of the bombadier's spray on insects and confirmed it was definitely defensive. Ants 'zapped' by the hot cloud stagger about in a very disoriented fashion but it does not normally kill them. The mechanism is used sparingly to give the beetle time to escape. So far as Eisner has been able to establish the spray is effective against most small predators with the exception of one species of spider, which has learnt to spin bombadiers in cocoons of silk that incapacitate the spray nozzle, and certain ants who will commit 'altruistic suicide' when attacking one. These ants keep up a repeated attack, many sacrificing their lives to the spraying, but eventually the bombadier's chemical reserves will be exhausted, and the remaining ants will overpower it.

Extraordinary as this jet propulsion chemical laboratory complete with rotating nozzle may seem, such systems are almost common in the inner universe of the insect. Eisner knows he has done little more than disturb the surface. Already he has found another beetle with bombadier-type chemical sprays that uses the adherent properties of liquids running

The master chemist of the insect microuniverse, the bombadier beetle. Eisner's remarkable picture shows a bombadier spraying through its rotatable nozzle a pulsed, noxious spray heated to boiling point.

along surfaces to bend its spray round corners. Certain millipedes produce droplets of a similar chemical, while other millipedes produce droplets so toxic they can stun large wolf spiders. The grasshopper, *Romalea microptera*, produces a repulsive chemical froth; the larvae of the sawfly, *Pseudoperga*, vomit oily fluids in the faces of potential predators and, most bizarre of all, the larvae of the beetle, *Hemisphaerota cynaea*, weave and hide inside a delicate string net of solidified faeces.

Recently Eisner has started to consider these defensive mechanisms in a new light – as a form of non-verbal communication. Sprays and other substances 'tell' attackers to go away, in the same way as silent displays of aggression warn off other animals.

Communication in the insect world is Eisner's other great passion, in particular a method of communication that insects have developed more successfully than any other species and which does not exist among humans – the use of ultraviolet light. The pioneer work in the field was conducted under the direction of the bee master, Karl von Fritsch, when it was demonstrated that certain plants were giving bees strong ultraviolet directions to nectar sources.

Tom Eisner has managed to build cameras based on commercially produced closed-circuit television equipment, which allow him a continuous view of the ultraviolet world of insects and plants.

Humans see light in the 400 to 700 Ångstrom waveband (blue to red). Insects see light at wavelengths of 300 to 400 Å (mainly blue). Previously it was thought that the only use being made by nature of the ultraviolet part of the spectrum was in some flowers that had an ultraviolet tone, giving them a special glow. Eisner's more sophisticated ultraviolet cameras have shown that many plants have distinct ultraviolet patterning, and a number of insects, particularly butterflies, 'wear' what appear to be ultraviolet 'badges' of identification. In normal daylight we cannot see these patterns, but when viewed through an ultraviolet filter, the effect is quite dramatic. Compare the two yellow butterflies above. When seen through an ultraviolet lens, two large white flashes 'light up' on the male. Similarly, the black pollen area of the black-

Now you see it. Insects use ultraviolet signals extensively (note the white area).

Now you don't. The white 'flag' on the wings is invisible in this normal light photograph.

eyed susan, which appears fairly small to the naked eye, becomes a massive bulls-eye under ultraviolet light – the way a pollinating insect would see it.

It would seem that most of nature's use of ultraviolet is related to procreation: insects signalling to potential mates, plants signalling to pollinating insects and so on. Two of Eisner's students have proved that this is the case with some butterflies: when the ultraviolet flashes on their wings are filtered out, they are quite unable to find a mate.

Eisner's ultraviolet work is still at a very early stage. He recognizes that he has done little more than lever open the door to a complex world of insect communication that we have previously only suspected might exist. But it may well be that before any further statements can be made about how insects communicate with each other, and how animals communicate with plants and vice versa, we will have to completely re-examine the whole of nature through an ultraviolet lens.

That is a life's work for several people. 'If anyone asks me' Eisner says, 'how much we know about insects, I feel obliged to answer "virtually nothing". Until about 50 years ago, most information about insects was of an observational nature. Now we've started to apply a lot of experimental work to our inquiries. But only a relatively few species have been observed in detail and there are millions of them.

'The insect world is really a great big unknown. It's the counterpart of outerspace. We spend millions travelling to the moon but all we have to do is look under a leaf to discover a whole world that is new.'

If you detect a small note of irritation in that reference to the NASA budget, you would be right. The American academic establishment is under some pressure at the present time to justify its existence, and pure researchers, like Tom Eisner whose work might not appear to be of immediate benefit to the human race are somewhat vulnerable. To the contrary, Eisner points out that his work in insect chemistry should be regarded as very important to the human race. After all, it is against insects that so many manufactured chemicals are directed and there is still very little information either

on their efficiency or, more importantly, on their harmful effects on the environment. And then there is all the chemical industrial waste dumped into the environment, the effects of which are often not even considered.

'We really don't understand nature in chemical terms yet. We don't know what a chemical adds to nature because we don't know what the chemical inter-relationships are in nature that might be affected by it.

'If I can unravel a tiny bit of nature's chemical structure, this illustrates how many more things are going on in nature of which we know nothing. Also it shows how foolish it is to tamper with nature without understanding what the long range effects are of this tampering.'

Dramatic ultraviolet beacons are transmitted to insects by some plants. This is the black-eyed susan photographed as we would see it, in normal light.

Eisner emphasizes strongly that to admit that he, a leading scientist in the field, knows next to nothing about the insect world can act

as a counter to the excesses of industry.

'All too often industry will say there are no known effects to a particular chemical, and hence it is safe to add it to nature. I'm saying we don't know and if we err we should err on the side of caution.'

He insists that the human race, in its own interests, should ask itself whether it really needs to add chemical preparations to nature's already delicate balance.

Fortunately for Tom Eisner (and in our opinion, for nature) it is unlikely that any governmental meddling will throw rich Cornell University's hallowed biology faculty off the trail of pure research. The place suits Eisner very well because essentially his interest in insects and their world is based on a fascination so deep rooted he finds it hard to justify and cannot understand why other people do not feel it as well.

The same plant (as left) under an ultraviolet lens. A pollinating insect is given a black bulls-eye to aim at.

'If you believe there is still in human beings a basic love for nature, then some benefit can be harnessed from our discoveries in terms of keeping that interest in nature alive.' At which point he threw up his hands in frustration and cried, 'God that's awful. How can you not be interested in an animal like the bombadier beetle! It's a fascinating beast! So, for that matter, is any animal or plant if you spend any time with it!

'Maybe it's essential that we do this work to encourage a little more humility on the part of humans and their relationship with the rest of the world. I do believe that it is the responsibility of anyone in biology to keep the conservation ethic alive.'

Unfortunately, in company with many naturalists, he is not too optimistic about conservation. Eisner himself serves on a number of conservation action groups, one of which was responsible for the rescue of the Texas 'Big Thicket'. A few miles from his home there is a section of virgin forest that

'Most people feel the world is our domain. We have to change that rather narrow view.'

Eisner is very fond of and where he sometimes goes ostensibly to collect insects but basically just to sit and think about nature.

'You have to travel 100 miles to find another piece like this,' Eisner will point out. 'It's a little pearl in an area of unnatural habitat. One of the things we're facing is habitat extinction; we've essentially decided that the world is there for us to conquer.

'If, however, you have the attitude that we're not necessarily the only things on this earth which have a right to exist, then your attitude towards habitat changes somewhat. But this may require a massive and persuasive effort. Most people still feel the world is our domain. We have to change that rather narrow view.

'My priorities are now very simple. I feel there is so much going on that is still unknown, it's our obligation to preserve the natural wilderness. I'm also convinced that everyone is interested in nature. All you have to do is to point out what's there and that it's there for everyone to explore. It's not expensive to explore it. You don't do damage to anything by exploring it and it will still be there for our descendants to share. They won't be reading about things that have become extinct – they'll be able to go out and look at things for themselves. If we were to decide that this is really what we leave to our children and our grandchildren, not edifices or pieces of engineering, that's the kind of legacy and tradition I wouldn't mind at all.'

Reference

ANESHANSLEY, D., EISNER, T., WIDOM, J. M. and WIDOM, B., Biochemistry at 100°C: the explosive discharge of bombadier beetles (Brachinus), *Science*, 165, 1969.

BLUM, M. S., Insect defensive secretions: hex-2-enal-1 in *Pelmatosilpha coriacea* (Blattaria) and its repellent value under nature conditions, *Annals of the Entomological Society of America*, 1964.

EISNER, T. and MEINWALD, J., Defensive secretions of arthropods, *Science*, 153, 1966.

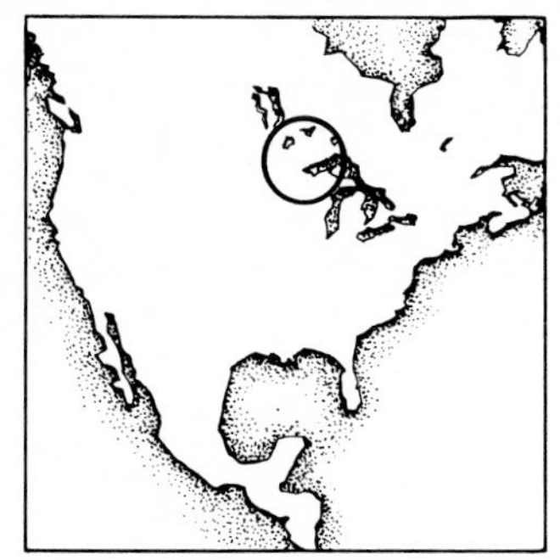

LYNN ROGERS

No Picnic for Teddy Bears

Kawashini Field Laboratory, Ely, Minnesota, USA

'If you go down in the woods today, you'd better not go alone. For every bear that ever there was, is gathered there for certain because. . . .'

Dr Lynn Rogers knows exactly where all the bears go for picnics – all those within 100 miles of his research centre near the town of Ely, Minnesota, that is. And it is not bears' picnics that interest Dr Rogers, but bears at human picnics. An important part of his study is to investigate what might be called the 'Yogi-syndrome' in the North American black bear – an absolutely insatiable appetite for human food that often costs the bears their lives.

Ely, Minnesota, sits just south of the Canadian border and is the tiny capital of a self-styled kingdom 'the land of 10,000 lakes'. A couple of hundred years ago the area was explored by those tough European hunter/explorers, the voyageurs. They were attracted by the numerous species of fur-bearing animals including otters, beaver, wolf and a great number of black bear.

The voyageurs also established another enduring tradition in this area, the persecution of wildlife. By 1840 they had shot and trapped many of the fur-bearers to the threshold of extinction. Fortunately, the lack of animals forced them to move further west and the few remaining breeding pairs made a rapid recovery in the empty wilderness. Today the voyageurs are back: Ely sells voyageurism in neat packets to a million tourists a year.

Lynn Rogers with the loves of his life, black bears.

White-tailed deer and black bear abound in the woods of northern Minnesota. Both species are the subject of Lynn Rogers' long-running habitat study.

For your vacation fortnight you get yourself 'outfitted' by one of the several firms in Ely specializing in complete, if somewhat plastic, voyageur kits and pretend you are braving the Frontier. Americans are very fond of this fantasy even though their excellent technology has removed them further from the primitive than any other human group. Hence today's voyageur canoes are aluminium and the trade is based on the fact that they have not lost a tourist in years, or only just enough to keep the land of 10,000 lakes vicarious.

It would all be completely innocent were it not for the fact that numbers of these ersatz voyageurs carry rifles, shotguns, pistols and lethal bows and arrows, even though their food is supplied in freeze-dried form.

The bears just love the food, particularly freeze-dried chocolate ice cream (which tastes like stale cocoa), but the price of hijacking one of these outdoor TV dinners is very often an inexpertly placed bullet or arrow.

Like the cartoons say, black bears do scrounge.

In trying to strike a balance between the tourists and the bears, Lynn Rogers has got himself caught in a trap of desperate irony. Bears are on the increase, due in part to the fact that human food, rich in fat and protein, breeds bigger bears. To quote friend Yogi (who's whole act is based on his adept scavenging), bears that can get their paws on human food will grow 'better than yer average bear!'

But the tourist trade is also growing – in fact booming – and as Ely merchants cannot afford a real bear threat, more and more bears are being shot; there can be no doubt that in the long run the land of 10,000 lakes is going to be over-run by people rather than bears.

It might seem from the large numbers of bears around the campsites, that the balance was holding level, but according to Lynn Rogers (although loath to admit it, because like so many other American naturalists, substantial amounts of his funds come from hunting groups), the Minnesota black bears are probably already losing.

It is quite extraordinary how many naturalists either physically resemble, or have taken on one or more of the characteristics of their subjects of study. Lynn Rogers is no exception: a very large man, bright eyed with a red beard, he walks with a slightly rambling, bear-like gait, and most strikingly, he grunts in a huffing, breathy way – very like the noise a bear makes as part of its aggression display.

Admittedly it is not that difficult to find resemblances between bears and men. In fact, they are among the most man-like of all animals, which accounts for the love-hate relationship we have always had with them.

'They're certainly the most man-like of the North American animals,' Lynn acknowledges. 'When they stand up they're almost

Big Richard, Rogers' favourite bear who learnt to sit and wait for his medical when caught in a research trap.

our size, they have nearly the same musculature and the same skeleton; they just lack a collar bone. And they like the same food.

'Their main diet is berries, but they can eat meat and they will eat vegetables, all of which are low in fat. But human food, which they can also eat, is very high in fat, as well as protein, so if they can supplement their carbohydrate berry diet with human food they grow faster, mature sooner and reproduce more successfully. They can become biggers bears and win more of the mating battles. I'm afraid there's every reason in the world for them to want human food, especially in years when the berry crop fails.'

So Goldilocks would have got an angry reaction from the bear family if indeed she had gobbled their porridge! Although generally just greedy, they can sometimes be vicious.

'It doesn't pay to mess with them. A male can get up to 600 lb and it's amazing how much power they've got in their jaws. They seem to build up such a mass of jaw muscle, that they can bite so hard they break their own jaws. I think it's a kind of evolutionary disorder in the males.

'They are not as aggressive as the grizzly bear, but black bears are occasionally very savage. I know of about 35 more or less unprovoked attacks by black bears, and 18 of those have been fatal. On the other hand, considering the amount of contact there is between black bears and people, the attacks must be considered very rare. In fact, there are a lot more encounters than people realize, but the bears just melt off into the bushes, avoiding an encounter whenever they can.

'And people behave very badly sometimes. A bear will come to a campsite and the people will throw food to it and try to entice it out where they can get a good picture, but the next day, when the bear is bolder, they don't like it so much any more – it's become alarming by then.'

That is the point at which a rifle can suddenly appear or a call be made to the forest warden. Either way another black bear dies for simply doing what comes naturally to bears. While the bear's reputation as a scavenger is accurate, and they can sometimes be nasty if provoked, Rogers suspects that other features of that reputation are pure fiction.

The bear's 'vicious' claws, for example, are almost never used to kill other mammals. Claws, and the large canine teeth which give the bear a savage reputation, are used almost exclusively for ripping logs apart in search of insects, particularly ants. Sharp, tightly curled claws are also good for tree climbing and useful for digging dens and excavating roots of nutritious plants, hornets' nests and anthills. In fact, Rogers' study has revealed that a free-ranging Minnesota black bear spends more than three-quarters of its foraging time in midsummer investigating sources of ants.

The sleepy nature of the black bear, satirized in many cartoons, is reasonably accurate, but not to be relied upon. Food is scarce in bearland and the animal works hard to find enough. If needs be, moreover, black bears are able to produce bursts of speed in excess of 25 mph.

In reality, the animal's lifestyle is entirely dictated by food. Like man, the black bear does not digest cellulose very well. When the berry crop fails, bears that are forced to live on 'greens' become severely emaciated, more are seen on garbage dumps and attacks on humans increase. Most of the food sources of the black bear are available only at certain times of the year or are too small and scattered to be gathered rapidly. 'The foods that are potentially abundant', Rogers explains, 'are dependent upon the annual vagaries of temperature and precipitation, with the result that over much of the range of the black bear there tends to be a surfeit of food in some years and absolute or relative shortages in others.'

The black bear is assisted in its foraging by a quite uncanny sense of smell, acute hearing and excellent colour vision. 'Bears can actually smell what kind of insects there are, and where they may be found in sufficient concentrations in logs. The reason you rarely see a bear in the woods is that they have hearing more sensitive than man's and broad, soft foot pads for moving quietly downwind to identify the source of unusual sounds.'

Lynn Rogers' study is particularly interesting because it has been running such a long time. Lynn came to this remote stretch of pine forest in 1969, and in all those 12 years he has never once been absolutely sure at the beginning of a year that there would be enough money to get his small team through. But he believes in the value of long-term research: 'It takes some bears eight years to come to maturity and produce cubs. Most wildlife research is too short to get to the conceptual level. The resulting management has to be based on speculation rather than fact.'

In those 12 years he has caught and examined more than 1,000 black bears of all shapes, sexes and sizes and he has fitted over 100 of those with radio collars. At one time

Lynn believed he had every female bear in the study area broadcasting information. He also has a pretty good idea of the territorial routes of adult males. If the Minnesota black bears had gone on a group picnic, Rogers would have known about it.

In fact they never do. As with most animals, territory and ranges are firmly established and Rogers has demonstrated conclusively that large groupings of black bears do not normally occur.

There is, however, one known exception: black bears will come to human picnics, or rather to those places set aside for the remains of our picnics – the garbage dumps. Such is the black bear's insatiable appetite for rich human food (an inexorable drive in that it has evolutionary advantages) they will scavenge rubbish tips together.

Otherwise the rules of movement and behaviour in the bear world are fixed and polite. The female black bear establishes a territory – a large one of two to three miles in diameter – in which she raises her young and defends them against other bears. 'Sometimes she'll venture out for short periods and then come back, especially during the mating season when she's laying scent trails for the male to follow. She also uses her territory for her young. At the age of a year or a year and a half when the young go their own way, each one of her cubs will grab a little bit of the mother's territory. The mother will avoid that place and leave it as the exclusive feeding area of the cub. We discovered that cubs which did not have this facility grew more slowly.'

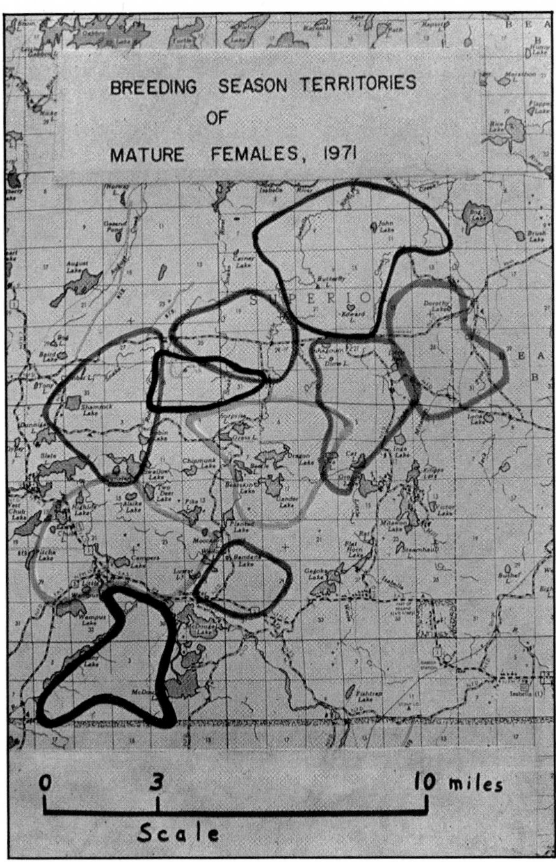

Female black bears are territorial, allocating some of the area they control to their cubs.

Despite the fact that 'voyageurs' to the area are warned not to come between a mother and her cub, Rogers has found that black bears are generally unprotective towards their cubs – at least when they are out of the den – 'We commonly chase black bear family groups to tree the cubs and ear-tag them. As yet, mothers have done no more than bluff [with a charge]. Even when cubs scream "Maa" with almost human voices, the mother usually stays away.'

Rogers believes that this lack of 'natural selection for defence of cubs' stems from the fact that black bear live in forests where there are plenty of trees into which they may escape. Other types of bear, particularly the other North American bear, the grizzly, do defend

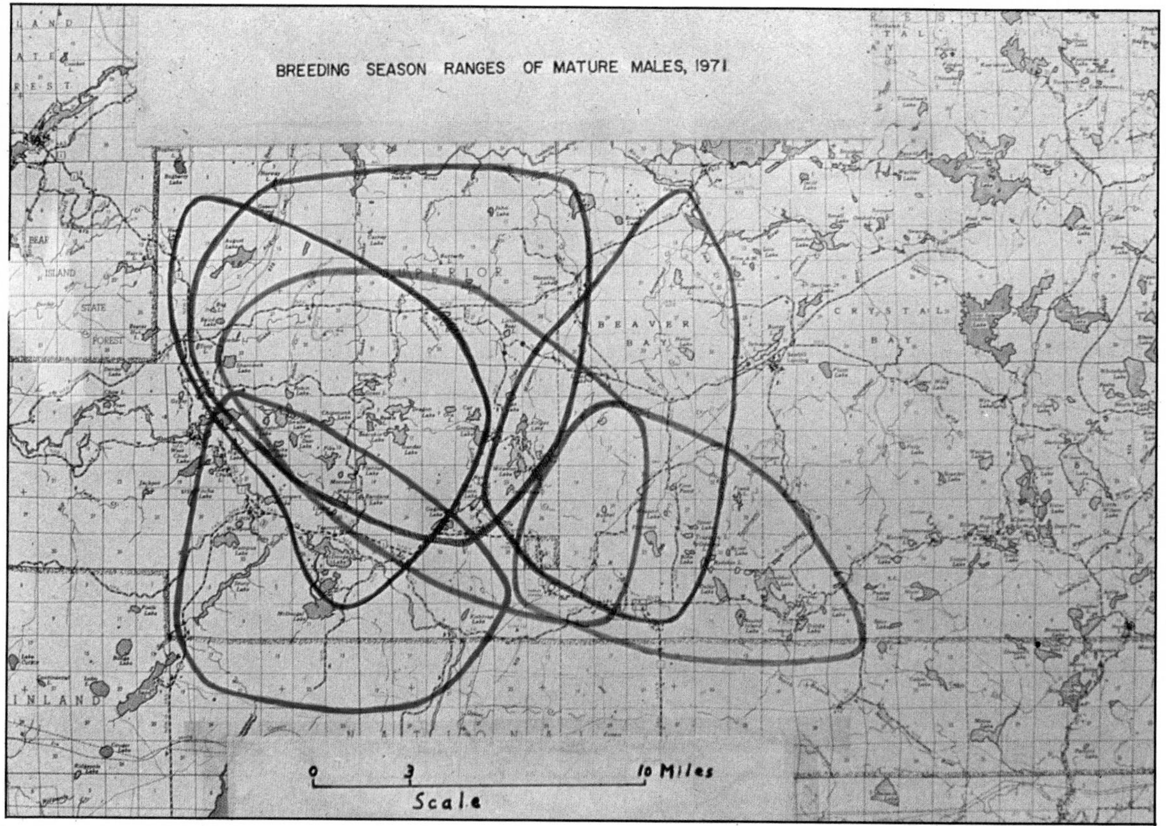

Male bears have ranges, rather than specific territories, and they cover a lot of ground.

their cubs aggressively, but they live in fairly open scrubland.

When the cubs are two, or sometimes three years old, the young male bears must wander off and establish their own ranges. 'They go far. It's often more than 100 miles before they can settle down. On the way they run into humans and human food and of course some of them get into trouble.'

Lynn's study has shown that if you time it right it is possible to move bears to more remote areas where they will encounter fewer humans and from which they will not return. This works best with immature males, who also make up the bulk of the 'nuisance bears', as they are called. If you try to move a mature female who is protecting a feeding territory for herself and her young, the chances are she will come back.

In fact, a bear's ability to find its way home very rapidly has surprised Lynn. 'One bear came back 142 miles to the exact spot from which it was moved. A bear in Michigan was moved 96 miles and 30 days later it was shot back at its original location.'

All this information comes from the radio collars Lynn and his assistants fit to the bears they are able to catch. The catching is somewhat messy and raises that question about naturalists which most of them would rather avoid: does anyone have the right to interfere with the normal life of a wild animal? The traditional answer is 'Yes, if it is in the animal's interest'. We have seen situations where this is hard to justify but not in the case of Lynn Rogers' bears. In an area where 90 per cent of adult bears die by the bullet, any work that will help to reduce this level of slaughter is worth a little indignity.

And Rogers has managed to convince the powerful hunting fraternity of the need for limitations. Until 1965 bears were bountied in Minnesota, and people were paid for shooting them. Even when the bounty was dropped, it was still open season on bears – they could be shot at any time of the year. 'A lot of people would go to the garbage dumps and shoot at rats. They'd see a bear and put a bullet in the bear too. There'd be a lot of bear carcases just rotting on the dumps in the summer.

'Then in 1971, based partly on the information from this study, the bear was made a game animal and the season restricted to the Fall when the pelt is prime and the bear would be most likely to be fully used.'

'Using' an animal fully means taking full advantage of the corpse in terms of meat, skin and trophies. Deer provide good meat and a coveted trophy, the head, but the skin is rarely taken. Bear meat is not often eaten but the thick black pelt makes an excellent trophy rug and heads are used for wall mounting (although this practice is decreasing because of the cost of taxidermy). As a result if you want a thick, glossy bearskin rug rather than a corpse that is not worth skinning it is only worth shooting a black bear in autumn.

The new hunting season has both advantages and disadvantages for the female bears, however. On the positive side, females are now not shot in the spring when they have

The one we would all like to cuddle – unwisely – a black bear cub.

cubs that would die as well; but as Rogers has discovered it is almost impossible to tell whether or not females killed in the autumn are pregnant. 'The denning period is the time when the bears give birth. Cubs are usually born in late January after a gestation period of several months. They are conceived in June or July, but development of the embryo is limited almost entirely to the last three months of gestation. Before that time, the fertilized egg is not implanted in the uterus and is barely visible without a microscope.'

However, Rogers does not entirely disapprove of hunting. Good conservation and habitat management can produce a wildlife boom that will need to be culled one way or another. In well-managed American wildlife habitats, hunters are used to carry out the cull and Rogers agrees with this method. Certainly he finds it a better alternative to landowners 'controlling' nuisance bears by the well-known and particularly unpleasant method of 'gut shooting'.

Bears are trapped for regular medicals in a ventilated tunnel of welded oil drums.

'You shoot the animal in the intestines so that it doesn't kill the bear on the spot. The bear leaves and the landowner doesn't have the bother of burying such a big heavy thing. In fact, one of our radio-collared bears was shot in that way. The bullet nicked the pancreas and over the next few months he managed to digest most of the blood while losing weight fast – nearly 40 lb. That animal took nearly five months to die.

'He finally went to three dens and when winter came, I could just imagine him trying to curl up in there with his big bloated stomach, being in pain; then leaving that den, trying another and finally finding a third. He couldn't handle the pain; he died wandering around in the snow, just skin and bone.'

A story to justify almost any indignities his study may cause the bears! A certain amount of indignity is inevitable when catching bears, because the animals are very big and very powerful and the traps need to be mighty as well. Rogers generally uses a barrel trap, which consists of two 45-gallon oil drums welded together. A strong metal trap door drops behind the bear when it disturbs a trigger. The traps are monitored daily, well ventilated and set deep in forest shade.

When we were there, Lynn received a report of a big male in one of the traps and we went with him to watch the animal being fitted with a radio collar. The team also traps bears to allow tagging, checks on weight, general medicals and measurements.

Even though Lynn has done this more than 1,000 times, he is always wary and approaches the trap with caution. It seemed impossible, but even big bears had managed to turn round in the barrel – and come out fighting.

Huffing more than normal, Lynn circled the barrel and then peeped through one of the ventilation holes. He was less than happy as it was a hot day and the bear could prove very irritable. Lifting the metal trap he injected tranquillizer into a black section of doormat, then shut the door quickly. 'It's difficult to know how much to give', he explained. 'The amount can vary from bear to bear and summer to winter, and, of course, we don't know the exact weight of the animal when it's in the trap.'

After about 10 minutes, the bear was hauled from the trap. It was very large, waved its sharp claws and displayed more than ample teeth, but as the tranquillizer took hold it settled on its side. Sharp and somewhat beady eyes still kept a good watch on all of us.

'Tell me if he curls his lip', Lynn requested. 'That's the first sign of aggression. At the moment he's quiet and peaceful, so we'll get on.' He took what seemed to be a lot of blood, but then there was a lot of bear. 'We're not just interested in learning how bears and men can better coexist', he explained, 'It has been discovered that the blood of bears in the winter is in the same condition as the blood of humans with certain diseases. These samples could give the researchers ideas for improving human medicine.' Further research, using blood samples from his bears, is being conducted to establish whether hibernation is triggered by a hormone.

Keeping a watchful eye on this lightly tranquillized bear, Rogers makes it a radio collar.

It took a long time to fit the radio collar. Bears' heads and necks are virtually the same width and the collar needs to be tight, without rubbing. The bear soon gets used to the collar, and if it stays on for a week, the animal will no longer notice it.

Once the collar had been fitted, the bear was measured, and a urine sample taken. Its head was being stroked rather absentmindedly when, abruptly, it sat up. We were all 10 feet away, galloping for open ground, when the bear flopped down again. Lynn went back to fetch his medical box and suggested we leave the bear to wake up in its own good time. In fact as we moved off, it got to its feet and wobbled off into the forest.

The study goes on all the year round and as this part of the world has a long and icy winter, the cold weather work can get very harsh indeed. In temperatures that often reach 30 degrees below, Lynn and an assistant will track a hibernating bear by radio, follow the line on snowmobiles until the snow gets too deep, then complete the journey to the den on snowshoes.

The amount of time a bear may spend in a den is dependent on that inexorable pendulum – food. In the north, black bears may spend as much as five to seven months in dens depending on local food supplies. In the south, where food is available for much of the year, bears den for only short periods or – surprisingly – not at all.

Once in its den, the bear's metabolism changes dramatically. As well as a marked slowing of the heart beat, breathing goes down to as low as one breath every 45 seconds, and kidney function almost to nothing.

Through Minnesota's harsh winter, Rogers and his team use snowmobiles to keep an eye on bears and cubs in remote dens.

But beware: a hibernating bear is not actually unconscious. 'They sleep in cycles and sometimes you catch them on an up stage and sometimes in the down stage. One I caught on the down stage! I wanted to get information on the heartbeat during hibernation. I crawled into the den, and as the bear showed no sign of stirring, I laid my head on its chest and listened for the heart beating. Nothing – in hibernation the heart rate can go down as low as 8 beats a minute. Then all of a sudden I did hear a very strong, fast heart beat – 150 to 200 beats a minute. Just at that moment the bear started to raise its head. I really shot out of there, but I swear I could still hear that heart beat. I even took my own beat to check it wasn't mine!'

So now, even in deep mid-winter, all bears in dens are approached with reverence and tranquillized with a syringe on the end of a long stick. Winter is an ideal time for renewing the batteries in the radio collars. The bears do not have to be trapped and are easy to find as they normally use the same dens each year.

In the spring, with the bears still half-asleep, Lynn returns to check up on any cubs. Mother and offspring are taken from the den, given a cursory medical and weighed, before being tucked in again. It seems remarkable that a hibernating mother is able to take care of a cub, but apparently she wakes instinctively when needed – a fact that Lynn almost found out the hard way.

He discovered a den with a big 'front door' and cubs inside making a lot of noise. 'I'd

Vulnerable cub? Not so. Rogers discovered that its alarm call rouses sleeping mother in seconds.

always wondered what stopped a predator taking the cubs from a sleeping mother. I checked the den and found the mother sleeping soundly with one of the cubs nursing. There was another lying quietly next to her and I really thought I could grab that one without her noticing. As I wanted to find out whether there was any alarm call that would wake the mother, I switched on my tape recorder, reached in and grabbed the cub quickly, just like a predator would.

'I got the cub out, and he gave a tiny little whimper that barely recorded on the tape. I started to put him back in, having convinced myself that the cubs *were* vulnerable to predators. In the second and a half that had elapsed between the time he made the whimper and the time I had put him back in the den, the mother had woken up and was lunging, her mouth open, towards the entrance. She inhibited her bite when she realized it was going to be her cub she was biting, but the momentum still knocked the cub out of my hand. He landed on his back inside the den but then he toddled over and climbed up on her back. She retreated to the back of the den and stood there blowing the top of her teeth – the sound bears make when angry. They don't growl, they sort of huff. . . .

'The snow starts to melt in the spring and the doors open up. Between that time and the time when the bears and the cubs emerge, there's practically no mortality, so I'm convinced the bears are almost invulnerable to predators when they are in their dens.'

One exception to that was a bear killed by wolves – there are still wolves in northern Minnesota and, not surprisingly, a considerable number of them are running around in radio collars. By chance a spotter plane was photographing wolves around a fresh kill that turned out to be a bear: a very particular bear that Lynn Rogers had been studying for 10 years. It was a very sad moment for him when 'Number 320' and her cubs fell to a pack of nine wolves in mid-February.

Apart from the wolves, only man is a serious predator on the bears. But as is so often the case, to have man on its trail is about the most dangerous situation any animal has to face.

No one is absolutely sure of the natural lifespan of a black bear. The oldest one that Lynn can recall was 21 when it was shot, and the unbelievable truth is that very few bears will see 20 years, because most of them will be shot long before they reach that age. In fact, 90 per cent of adult bear deaths are caused by gunshot wounds. Moreover, most of these bears are shot while attempting to obtain garbage near human habitation. Rogers' survey showed that such mortality was highest during years when natural food was scarce. 'In seven years of study in Minnesota, nine radioed bears were killed as nuisances during three years of scarce natural food, and only three were killed as nuisances during four years of moderately or exceptionally abundant natural food.'

This is why Lynn's radio tracking is so important. Other census methods are very crude, and without reliable and immediate information on actual numbers of animals, the bear population could be halved before anyone realized.

Yet, so far as we were able to establish, the bears were not doing anyone any real harm. Unlike other large mammals they make very little impact on their habitat and certainly do not destroy it. Lynn agrees that hunting is not needed to keep the bears from over-riding their habitat. Instead it seems that all the killing was permitted as a means of recreation and of preventing nuisance activity.

This Minnesota story is a classic case of the pointless way man will exploit nature. Here is an area of wilderness where black bears flourish and have achieved a natural balance with the habitat and their animal neighbours. We have invaded the area in the interests of our own recreation, and not satisfied with that, have accorded ourselves the right to shoot the bears for fun or if they act as nuisances, even though the source of the nuisance is of our own making: the garbage which we cannot be bothered to carry away or bury properly.

But the exploitation does not stop at the bears. There is also a deer hunting season in Minnesota. Aside from his studies of black bears, Lynn Rogers is also conducting a deer survey, itself the product of a rather bizarre feature of American wildlife law. The wolves of northern Minnesota are a threatened species, and the timber companies and forest managers are obliged by law to ensure they do nothing to the habitat that would be detrimental to the wolves. As deer are food for wolves, this has meant that they also have to be careful about deer habitats; and so Lynn Rogers received the money for his deer survey. Many people believe that men and wolves are the finest predators on earth: Lynn is in the position to study the prey of both.

His deer survey has been conducted with incredible thoroughness, notes being made of every mouthful of food that a deer eats all year round! There were three traditional ways to study feeding habits: to find out where deer had been feeding and note what was missing, to examine the stomach contents of dead deer, or to put a deer on a lead and note what it ate. Lynn disliked all three methods. He acquired eight young deer, just one day old, and hand-raised them. They became imprinted on him and he became what he terms 'an honorary member of the herd'. As such he, and his team, were able to stay right among the small herd as they grazed naturally.

Ironic fundraiser for Rogers' deer-study, the wolf. A threatened species, their food (deer) is protected by US law.

At first, however, Rogers had a problem: 'I noticed that some days the deer would skitter off and I couldn't get anywhere near them. It was just as if I'd developed a problem of the sort your best friend can't tell you about.' In fact he had B.O. – bear odour – and no matter how well he scrubbed himself after a day tagging bears, the deer could invariably pick up the scent. As a result a gap of several days between bear contact and deer contact became the rule.

After that their lives became pure Walt Disney (indeed the deer were of the spotted 'Bambi' variety), and during these forest wanderings Lynn developed a great affection for the gentle, beautiful animals. Unlike the bears, which had numbers, the deer were given names. 'We put hundreds and hundreds of hours of time into those deer. At the end of the second winter one of our animals, Bowser, became a leader of a herd of wild deer. Eventually the wild deer got used to us and they would lie down, with their fawns, near to us. We could sit down and watch tame deer and wild deer feeding all around us.'

It must have been a magical experience, to be so close to nature in the beautiful spring woods by the tranquil lakes.

During part of the year, the deer were watched through a 24-hour cycle with the team sleeping out in the forest. They shared watches so that someone was able to note the deers' activity at all times. 'It's the best job I ever did', Lynn says with real enthusiasm. 'You sit there learning the secrets of nature that no one has discovered before, seeing patterns develop in the deer, and writing it all down. It gives you a feeling that you are doing something worthwhile.'

And indeed he has made some very considerable discoveries of great help to his sponsors. Research into historical records revealed that the deer were occupying the area *thanks* to man. 'Around the turn of the century, people came to Minnesota, logged the pines and set off big forest fires that changed the habitat drastically. It opened up the land which then grew more scrub that provided food for deer. But now that this forest is growing mature again it's deteriorating as a deer habitat. If we want to keep the deer here we will have to start planting suitable food bushes and trees for them, and to do that we need to know exactly what they eat and when.'

There can be no doubt about the value of Lynn's deer study. It seems to be in everyone's interest, even the deer hunter's. But, unfortunately, even a deer wearing a very distinct radio collar is not safe in this neck of the woods.

Four of the original hand-raised deer have been shot, including the famous Bowser. For the first weekend of the hunting season, Lynn and his team walk out with the four tame animals they have left: 'whistling and calling to tell the hunters there is something unusual here. But it's hard to stay with the animals all the time and they do get shot.' The radio collars, clearly visible through the telescopic sights of his gun, are still no deterrent to the ardent hunter.

'In one case a man saw a deer with a collar and noticed that it wasn't running away like other deer. He got back into his car and actually drove into town, where he went to the Conservation Department and asked about it. They told him about us but also that it wasn't technically illegal to shoot the collared deer. So he drove back out to where the deer was grazing and shot it.' Later the hunter brought the collar back to Lynn who was obliged to thank him for returning it. . . .

It would be wrong to suggest that the wildlife in Minnesota is any more subject to the trivial whims of man than elsewhere in the world. But its exploitation there does seem a particular tragedy because the bear and the

deer exist in everyone's interest. A bear in the woods makes voyageuring that much more exciting, and the dappled deer, with their gentle grace, are very appealing. The land of 10,000 lakes is an ideal human recreational area and that side of its management is immaculate. It is surely time we recognized that the animals have at least an equal right to be there. At the moment the best ambassador they have is Lynn Rogers and we certainly hope his long-running study continues.

Imprinted wild deer allow Rogers to study every bite they eat. But in spite of their radio collars, they still get shot.

Reference

JONKEL, C. J. and COWAN, I.McT., The black bear in the Spruce-fir forest, *Widlife Monographs* 27, 1971.

ROGERS, L. L., Black bears of Minnesota, *Naturalist*, 21, 1970.

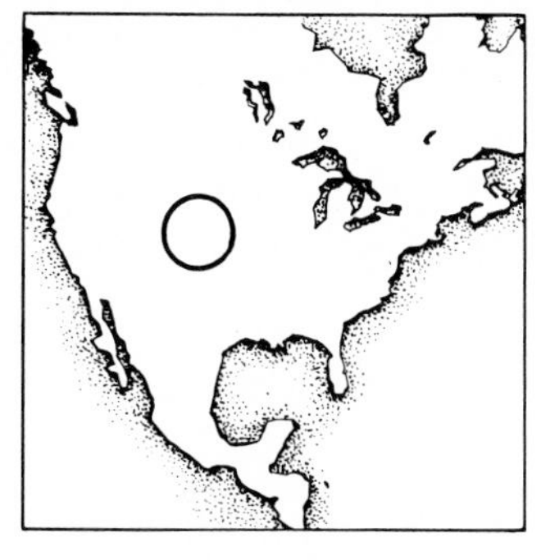

HARRY HARJU

A Licence to Hunt but not to Destroy

Cheyenne, Wyoming, USA

At the entrance to the Serengeti National Park, which is surely one of the world's great wildlife habitats, there used to be a sign reading 'Here the world is young and tender, kept in trust for your children and mine'. It was there 10 years ago and we hope it is still there today, because it seems to stress with delicacy the rarity and the relevance of virgin wildernesses of which all too few are left.

In particular there are very few left in the countries of the developed West. America is perhaps the exception. The United States are so big and the population so thin by comparison with Europe, that large sections of North America can still be considered virgin.

The state of Wyoming is one of them, perhaps the most spectacular. This single state is larger than England but has a population of less than a million. Wyoming is pure 'if only' wilderness. It is what Europe would have been like 'if only' there had been no industrial revolution; its high mountain forests are what the forests of Britain would be like 'if only' we had not wanted to rule the waves; its rivers are what the Rhine could be 'if only' industrial pollution did not exist; its wildlife is pretty much what the wildlife of the whole 'civilized' West would have been 'if only' early Man had not been such a very fine predator and his descendants had not considered that aspect of his nature such a fine tradition. . . . Well, almost!

Not very long ago, substantial quantities of coal were spotted in Wyoming. Even more recently, deep in the shifting mirages of the

Harry Harju, champion of Wyoming wildife, works on the principle 'compromise too much and you lose everything.'

Snowy Mountain spring. A corner of the vast nature reserve that is the state of Wyoming, now threatened by fossil fuel exploitation.

Red Desert where Pronghorn antelope and prairie dogs abound, uranium was found in massive quantities. Then, in the Grand Teton range near Yellowstone and up in the Snowies, where the aspens shake their silver leaves over big herds of elk and moose, a drill has hit natural gas. Even good, thick, black oil has been struck in a mountain meadow where five different species of wild flower create several miles of magic carpet every spring. . . .

Wyoming, where the world is still remarkably young and tender, is being pushed over the 'if only' edge by the fossil fuel crisis. The Arabs have crimped the great American umbilical – the fuel line from the gas station – and the people are panicking. Life without those gross motorized hip-baths, the Chevvies, Olds and Fords is seemingly inconceivable, and this nation of otherwise sane, polite gentle-mannered people is prepared to allow the rape of virgin, beautiful Wyoming so that it may continue to gobble an excessive share of the planet's organic power reserves.

No place on earth quite so needs a knight in shining armour to protect it from the fire-breathing dragons of the 20th century, and as Wyoming's hero would most likely live either in Laramie or Cheyenne, it is somewhat fitting that he should wear a Stetson and cowboy boots. Enter Harry Harju, more often than not out of a dusty sunset. For reasons that will soon become evident, Harry works long hours.

For the record, Dr Harry Harju PhD is the Wyoming Game and Fish Commission's Staff Game Biologist, although the more romantic description of his job is rather closer to the truth. Harry Harju stands between the animals of Wyoming in their uniquely unspoilt habitat and a mining industry now operating from a popular public platform that would like to strip the top soil off the state and gut it of raw fuel.

He is one man attempting to stand in the way of forces that in Europe a century ago created the Ruhr, the South Wales coal fields, the Potteries, the Midlands Black Country, and the open-cast monster-mines in Minnesota and Pennsylvania.

Harry Harju, the man and his problem, are what real nature conservation is all about in this pragmatic age, and it is not exactly gentle reading. Indeed, if this were the traditional type of natural history book, he should not be here at all because his deep freeze is packed with elk, moose, fish and other meats that fell to his gun or rod.

Harry stripped us of any romantic notions about himself and his job at our first meeting. His department is financed from game licences. He hunts himself; he told us about the contents of his deep freeze. 'But don't think that's the only reason I hunt. I like to eat game meat, and we have a harvestable excess here. But mostly I hunt because I like to hunt!'

On the wall hangs his Masters degree in Biology and Wildlife Management. Accepting that Westerners do wear cowboy boots, imply that they are all Wyatt Earp's great grandson, and talk straight from the shoulder, Dr Harju is still something of a surprise.

He does not think much of academic research and he does not like hypocrisy: 'this department uses more fuel than any other in the state; we're the gas hogs in these parts'. But he patently likes Wyoming and in spite of our severe reservations about his hunting, we accepted an offer of a week's tour of one corner of his huge state.

Having loaded camping equipment into his Game and Fish Commission truck and topped up his auxiliary gas tank with an extra 25 gallons, Harry told us we had no need to worry about food because he planned to feed us – on elkburgers. It seems we were about to see (and consume) some of the Wyoming wildlife reality, as he saw it.

Until then we had rather avoided thinking about the fact that a great number of Americans, including a large number of natural history scientists, see no paradox in limited hunting of wild game – the so-called 'harvestable excess'. It is true that a similar paradox exists between naturalists and 'field' sportsmen in Britain and other parts of Europe. Fishermen, wildfowlers and even fox hunters vociferously defend their role as conservationists and are probably responsible for the protection of game habitats that would otherwise be redeveloped or polluted. But rarely will you find natural history scientists among their ranks, apart perhaps from the fishermen.

Harry Harju is an open, honest man who understands the European naturalists objection to hunting. He has given more thought than most to the rights and wrongs of the issue, and can explain convincingly why hunting should exist in Wyoming and that it is a good thing, in fact, for Wyoming wildlife. As this question will continue to crop up in natural history arguments for many years to come, it seems a good idea to give Harju's argument space as a kind of hunting charter.

'A lot of people who are antihunting are really antihunter; they simply don't like hunters because they're such slobs – and a lot of them are. We have a lot of signs shot up, fences driven over, cows and sheep shot. Not all hunters are like that – in fact I don't call those kind of people hunters: they're just a menace, and we do everything we can to police them out.'

Elkburger supper: conservation by Harju's department produces a huge 'harvestable surplus' of game.

This is a very astute analysis of the game slaughter problem that exists throughout America and has been responsible for pushing a number of species, notably the bison, over the brink of extinction and is threatening contemporary species, like the bald eagle, with the same fate.

It is not responsible sportsmen with a vested interest in conservation, shooting on licences issued by Game Departments, that pose this threat. The problem lies with America's ludicrous gun laws, and the bizarre biology of a substantial part of the male population, which seems to depend for its masculinity on the number and calibre of guns owned.

In frontier Wyoming where the cowboy is a near-extinct species you can still buy a hand gun without any form of licence, the gun lobby is still a ferociously powerful political cartel, and no animal – or for that matter road sign – is safe.

Harry Harju loathes these unlicensed killers rather more than the average animal lover. They blacken the reputation of the outdoor people to whom he issues hunting licences, people who make a practical and direct economic contribution to the cause of wildlife management and conservation.

'Take a look at the facts about our game populations. When my department started up in the 1900's there were about 5,000 Pronghorn. We now have about 400,000. There were between 5,000 and 10,000 elk and they were all in Yellowstone. Since that time we've transported them all over the state – in fact, all over the Western and Eastern United States and even to some Asian countries. Wyoming alone now has at least 100,000 elk.'

These increases have all happened in spite

So prolific are Wyoming elk that animals are actually exported to other states, even to some Asian countries.

Flowers all the way to the horizon may soon be lost to open coal seams.

of the hunting licences issued by Harry's department. On the contrary, in Harry's opinion, they have happened *because* of the hunting. Conservation of animals in Wyoming is paid for by hunters.

'They have a vested interest. We are solely supported by hunting and fishing licences and an excise tax on hunting and fishing equipment. We don't get a nickel from anyone else. We don't get any legislative money.' This is particularly important to Harry Harju as it gives him political independence from a legislature dominated by mining interests; interests that he is fighting for game habitats.

'There are a lot of people who have a great affection for wildlife, but they don't give any money to help keep it, and it costs money. Out here it's the hunter and the fisherman who pay to look after the animals.'

Although we had all eaten elkburgers, and agreed they were fine eating, we also generally agreed that we could not be hunters. Harry smiled and said that if we ate meat we were really being hypocritical. 'Our efforts here have produced a game surplus. By setting up wildlife sanctuaries and wilderness reserves and state parks we've created protected environments in which the animals can breed much more successfully than in the original wild. If those surpluses are left unculled, you will have dead animals; habitats can only support a finite number of game. Surely, it is more humane to harvest the surplus and eat it. People tell me I'm cruel to shoot animals, then they go down to the supermarket and pick up a plastic packet that used to be an animal. I care about the animals I eat much more than that.'

By now we were becoming a little confused. Caring was shooting? The altruistic bullet was a hard one to bite! Harry grinned, 'We have very strict hunting controls here in Wyoming. When conservation efforts create a surplus, as they will inevitably if you have the habitat, you have a choice – quality or quantity hunting. Colorado, to the south, has opted for quantity hunting. Anybody can buy a licence over the counter and hunt all but a few restricted species, for a short season. We've decided on quality hunting.'

What this means is that the Game and Fish department very carefully adjusts the number of licences and the length of season to ensure that kills fall well within the 'harvestable surplus'. But before you make the mistake of thinking that this is a secret way of controlling hunting, it is not. It ensures what Harry calls 'high hunter success rates'. Most people kill an animal of the type for which they have bought a licence, whereas in the Colorado free-for-all there is a good chance you will not get anything (and, sometimes, may not make it home yourself). Harry likes his licensed hunters to fill their bag: his logic being that it encourages the skilled hunters, rather than the 'slobs' to come to Wyoming.

'There's one very special law concerned with hunting and game animals in this state. The best one we have.' Harry said. 'It's a very serious felony to waste game meat.'

After four days in Wyoming with Harry Harju we were convinced by his argument, although it is doubtful whether any of us could, or would, actually pick up a rifle and bring down an elk.

We think that Harju can demonstrate that the naked ape, at his present stage of evolution, is still much more a part of the jungle than he is prepared to admit, and that the sensible way of managing wildlife is to accommodate this reality. Most of us still do eat meat, a great number of us still do like to own a gun; we are still the most efficient predators on Earth. These are not features of human nature that can be bred out in a few generations, if indeed they can be bred out at all, because without animal protein fuel, man would be a different beast.

Harry Harju, who has the luck to live in a place that is still close to the Garden of Eden, has maintained contact with the essential ethic of the present naked ape species and adjusted his wildlife politics to it. 'People are sometimes shocked that I'm a naturalist *and* involved in the business of killing animals. Yet a farmer kills animals constantly to feed his and other families. When America was mostly rural that was true of everybody and you never heard of people being upset by hunting. Meat doesn't come in little plastic packages, but people can't understand that: they are so removed from it now they seem to have the opinion that it's unnatural to die.

'But when you live in country like this

Although big game hunting country, Wyoming supports a fantastic variety of wildlife. In particular the more arid regions accommodate many small species.

[Wyoming] and you see people hunting every Fall, and you hunt yourself, you get a little more realistic about death. You realize that man has always been a predator, that other animals have always died to keep him going and it's unnatural for a human being to be a vegetarian. If you look at your teeth you will discover you are an omnivore.'

Summarized, it seems that Harry Harju has made a good honest case for his wildlife morality if you face the fact that man does not always practise what he likes to preach.

But there are other excesses for which this human bull in nature's china shop can be blamed. Harry Harju believes that the real threat to Wyoming's wildlife is not his licensed hunters but the whole of human society and its totally selfish approach to nature.

'If I have to point a finger at the real wildlife killers I would have to point at people who are developing this country, the people who demand suburban homes, people who drive up to the gas pump every day and refuse to conserve fuel, the people who scream at County Commissioners to build them new roads. But most of all I'd point to industrialists who can't see anything but making a buck!'

Harju has cause for this outburst. For the last decade, since the Arabs discovered the oil weapon, the vast and mostly untapped fossil fuel reserves of Wyoming have been exciting speculative interest like a 20th century version of the Klondyke gold rush. The state has immense coal reserves so close to the surface they can be cheaply mined by that process most damaging to wildlife habitats – strip, or open-cast, excavation. It is also possible to scrape up yellow-cake uranium in giant buckets as fuel for nuclear power stations. Almost everywhere, even in the most beautiful mountain ranges of the Grand Tetons and the Snowies, drills are probing for natural gas

Urban sprawl around the mines begins to encroach on deer pastures and antelope migration routes.

and oil. Very few are coming up dry.

To all these facilities new roads have to be built and sewage and power facilities provided. Laramie is a boom town of mobile homes and only slightly more permanent, ticky-tacky houses. As with all boom towns, growth has been too quick, or too important, for town planning. And it is unlikely that these will ever become ghost towns. Everyone accepts that the substrata of Wyoming are almost pure fuel. The huge and almost indescribably ugly open pits of the existing strip mines have been operating for nearly half a century and show no sign of petering out, even though they are now deeper than the surrounding canyons. Only the animals that once grazed the scrub and grass on this neolunar landscape are ghosts.

In the 1960s Wyoming's tiny Game and Fish Commission became alarmed about the potential impact of an energy boom. They teamed up with the US Fish and Wildlife Service in a project designed to identify the relationship between land use and the distributions of elk, deer and antelope.

This survey is still going on but early counts show that the fears were more than justified. Throughout the 1970s several Wyoming towns saw massive increases in population, 50 per cent in Gillette and Rawlins, 90 per cent in Rock Springs while Douglas and Green River increased by more than 100 per cent. 'Potentially serious conflict between big game habitats and subdivision activity (residential sites and mobile home parks) exists in several counties, particularly Sublette, Park and Teton,' an interim report stated. It also reported that by the end of the 1970s there were nearly 30 mining operations in Wyoming 'that expect to affect at least 1,000 acres each', and large numbers of gas and oil fields had been discovered between 1967 and 1977, particularly in Campbell and Sweetwater counties, both important wildlife areas.

Harry Harju and two young assistants are charged with the awesome responsibility of attempting to insist that this industrial avalanche has a human face or, more accurately, an animal face.

Thanks largely to the personality of the man and an essential pragmatism that is perhaps reflected by his attitude towards hunting, Harry Harju, extraordinary as it may seem, appears to be winning.

First, he has heavily publicized the fact that

Wyoming's wildlife is a commercial resource in its own right, and one that could outlast the mines. He has estimated that the 'consumptive uses' of wildlife contribute well over US $100 million to the state coffers in a year, counting licence fees for hunting and money spent by visitors on motels, food, souvenirs and clothing.

'Mining critical wildlife habitats', he argues persuasively, 'is a bit like mining an area containing two minerals, extracting only one, and piling so much spoil on the area that the second cannot be mined economically.' It is a plausible argument and one that would undoubtably have prevailed years ago when American companies had firm control of Middle-Eastern oil. Today it could be just too intellectual.

At present the coal is coming out of the mines around Hanna as never before and the scar that is the mine and its peripherals takes up probably 10 square miles of Wyoming. In the Shirley basin, the open-cast uranium mine is now several hundred feet deep. Although the nuclear accident at Three Mile Island has caused a temporary hiccup in the US nuclear electric programme no one, in particular Harry Harju, thinks it will last longer than the next major increase in gasolene prices. Both mines are situated in very arid parts of Wyoming, have been there for some time and are almost impossible to reclaim. It is difficult to sustain vegetation growth under normal circumstances, almost impossible when the top soil that will eventually be put back has been piled in huge dumps which effectively sterilizes all but the top few feet.

Coal mining is now on the increase in the Powder River Basin. It is the centre of the largest concentration of Pronghorn antelope

Monsters of the nuclear age claw down in search of coal and uranium.

in the world. Bentonite mining has started on the west slope of the Big Horn mountains – a vital wintering range for mule deer and elk. There are at least a dozen applications waiting approval for uranium mines in the Red Desert. The desert supports the last really migratory herd of antelope in Wyoming.

Harju is a realist and he knows that the best he can do is fight a long-term rearguard action; a kind of guerrilla war with delaying tactics and red-tape as virtually his only weapons. But he has already managed – and it is a major achievement – to get the basic existence of a vital wildlife heritage recognized.

'There's a comment I've heard and read between the lines in many Environmental Impact Statements that says roughly, "Why worry about a few animals? We've got plenty of them". I would counter that by saying, "Why mine or apply for new leases on coal, or uranium or gypsum? There are plenty of other companies mining already'. The answer is that there is a demand for the resource, and mining companies make money by supplying a needed commodity. Well wildlife is a renewable resource, and there is a demand for that resource, a very high demand. But it is not renewable without habitat, just as a mine doesn't exist without a mineral deposit.'

If this sounds uncompromising, Harry has also used his department's well-read magazine, *Wyoming Wildlife*, to promote the same determined line. 'Although most people don't realize it, literally every development ever proposed has been detrimental to wildlife. As you read this your house is sitting on what used to be a wildlife habitat. Maybe only a few small mammals and song birds were affected, but the impact was there just the same.

'Those of you who are now saying "who cares about a few mice and tweety-birds" should stop to consider that passenger pigeons once roosted by the millions where there are now farms and cities in southern Michigan. Bison and elk were once found throughout the United States. The list of 'used to bes' is almost endless. Closer to home, there are people in the Cheyenne Game and Fish Department Office who once hunted antelope where there are now dozens of houses north of Cheyenne. The antelope are no longer there.'

Wyoming, Harry warned his readers, possessed a unique animal in the Pronghorn antelope: the state was home to more than half the world's population.

Of course, when the latest American fuel crisis reached its emotional peak with the taking of hostages in Iran, Harry recognized that he alone could not resist the pressures of the nation for fuel self-sufficiency.

Mining in Wyoming was as certain as tomorrow's cloudless blue sky. How could its impact on the wildlife and the ecology be minimized? The answer was rigorous implementation of the Environmental Impact reports that every mining applicant is required to complete under state law. Under Harju's supervision these have now become so detailed as to require the large mining companies to hire ecologists and biologists either full-time or on consultancies that may run for at least a year.

It is now an accepted principle that all habitats will be effectively restored after mining. And this does not simply mean filling the hole and planting grass. 'For wildlife, diversity in reclamation is the key word. We see this as meaning diversity in both topography and plant species composition.' More simply, if you want to mine Harju wildlife territory be prepared to do costly landscaping and complex gardening in the future.

He is also a vigorous opponent of uncontrolled, and largely unplanned, ribbon development of human facilities to service the new mines. In some ways Harju regards this as the more damaging, first because even the most enlightened ecologists rarely recognize that their homes (and their research stations) are taking up a former wildlife habitat, and second 'because this loss of wildlife habitat is likely to be irreversible. Even if the mine disappears in years to come, the roads and the home sites won't.'

Harju has also observed what he terms 'ribbon development' of mines and their associated facilities. 'Each development project chops off a little more wildlife habitat, even though each developer, mineral or otherwise, feels his project will have "little impact on the wildlife". Put together a whole bunch of little impacts on wildlife and you get a lot of impact on wildlife.

'Put one mine employing 100 miners in an area, and the impact on wildlife may not be large. However, put six to eight mines, each employing 100 miners in the same general area and you have a large impact on everything, including wildlife. Put that number of

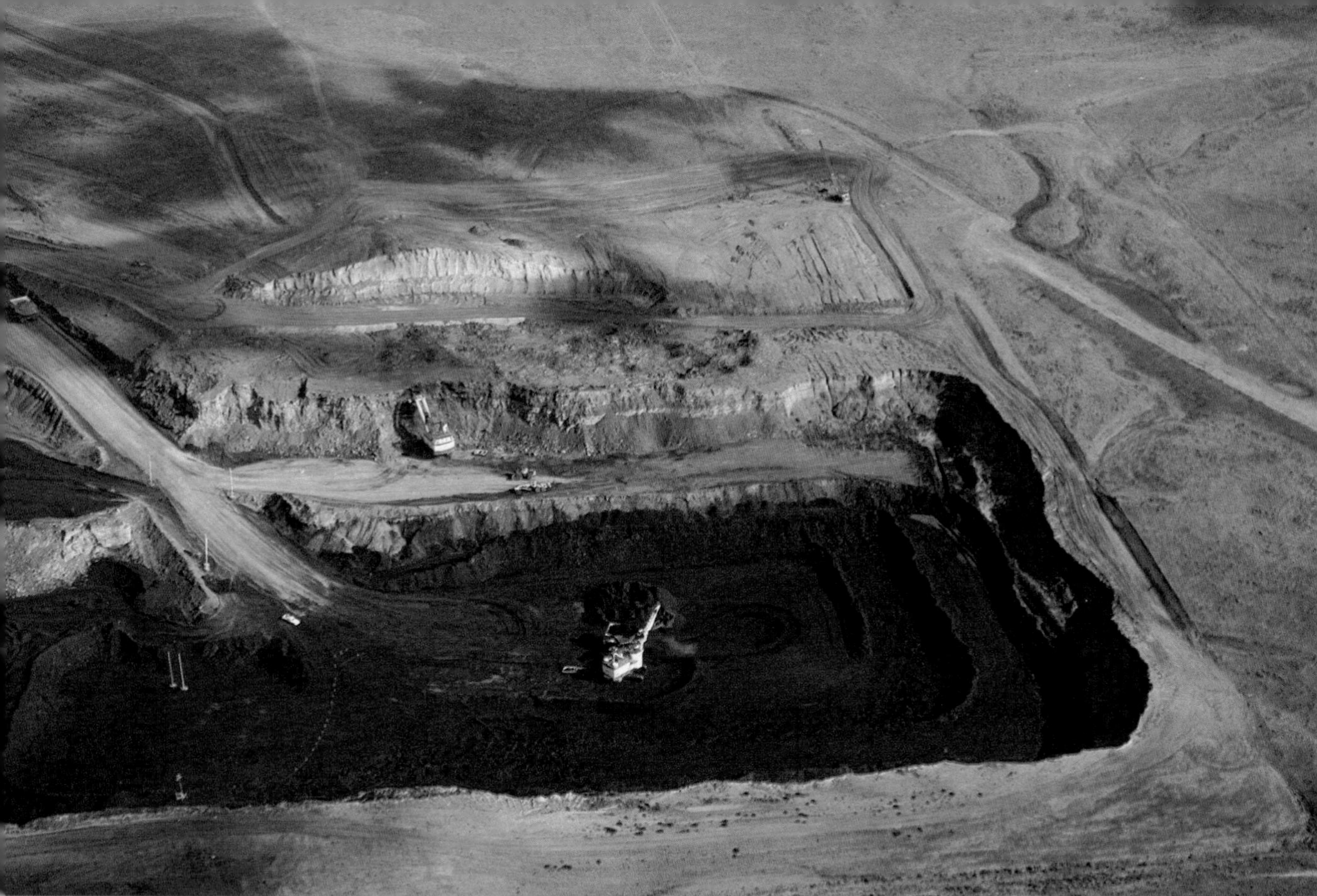

The black hole into which Wyoming's unique wildlife heritage could fall – an open-cast coal mine.

mines employing even more miners in the same general area and you have a situation similar to Rock Springs, Green River or Gillette.

'Associated with each new mine are land disturbance, fences, new roads, railway spurs, power lines, increased traffic, more demand for hunting, poaching – to name just a few wildlife-related problems.

'There are few benefits to wildlife from mining. Reclamation is not such a certainty that we are overly optimistic about the fate of mined wildlife habitat.'

Harju is attempting to insist that every human facility built in Wyoming should be reviewed for potential ecological damage – a unique concept if you apply it as widely as Harju does. Do not build a road unless you have to; do not widen an existing road unless you have to; do not allow urban sprawl and ribbon development such as is happening through neglect around Cheyenne and Laramie; control the trailer parks that spring up like fungus growths round the drills and embryonic mines, later to have their wheels removed and cement walls added. Power lines are unsightly and fences are a positive menace.

Harry Harju feels about fences as did the old Western cattle barons. But he has seen some ugly tragedies caused by fences that take no account of wildlife herds. 'Antelope don't jump over things, they jump across, so a sheep-tight fence can be lethal. In the Red Desert west of Orleans we found 75 per cent of a herd – 3,000 to 4,000 animals – dead in a fence one winter. They couldn't get through because of the piled-up snow drifts; they died of pneumonia.'

Driving with Harju in his big green truck is a nerve-wracking experience even though he is an exceptional driver. He talks a great deal, gives the impression that you have said something to annoy him and quite obviously does not have his mind on the road: not in the accepted sense at least – he sees open country. 'Twenty thousand acres of Wyoming are now under Interstate 80 [a main highway]', he said with gritted teeth 'and that's a damned sight worse than any mine. It's permanent. Mining land you can usually reclaim.'

Even accepting that Harry Harju is a very tough individual it was impossible to escape the conclusion that he might be attempting the impossible. Could he seriously contemplate a future for Wyoming completely removed from the realities of industrial progress. He believes he can, provided he can motivate people.

He took us to see Battle mountain near the town of Baggs (population 115). We suspect that although by now we were eating his elkburgers without flinching he was still uncertain whether we were properly motivated in other respects.

This beautiful peak, cloaked in dark pine, lifts out of a grassy valley thick with wild flowers and the shifting silver-leafed aspen trees. Rabbits and hares eat by the sides of the roads as though starring in a Walt Disney movie, and herds of mule deer roam the woods. The mountain is a home for eagles, moose and elk. Around Battle mountain at least 100 different species of game – nearly a third of the Wyoming contingent – live out their lives. Even by Wyoming standards it is an area of breathtaking natural beauty, much used for holiday and weekend recreation.

Harju hates fences because they limit the movement of his animals and can be lethal traps in winter.

Harry took us there in the interests of motivation. He had just seen an application for a coal mining lease covering an area of 30 square miles that would extend out from under Battle mountain itself, along the forest boundary into the mountain scrub habitat. He had made his point. What could be done against commercial pressures so philistine? A strip mine application for Battle mountain was like a proposal for the Garden of Eden.

Surely none of Harry's arguments could prevail against that kind of insensitivity. But Harry only grinned. 'We require that they survey the whole area very thoroughly, and there's a lot to survey here. I'm going to insist they do a seasonal wildlife survey, a check on what wildlife species there are, what time of the year they spend here, what habitats are important to them, and so on.

'Then they'll be obliged to restore the place. I think it's fairly obvious from looking at that mountain that it's going to be damned near impossible for them to restore something like that.'

Although Harry was not admitting to it, it seemed to us his strategy was to make it so difficult for the mining companies that they would decide Battle mountain was not worth the battle. Harry, straight as ever, said that was not the case. 'I'd like to, but I don't think I have that power. If they're going to mine the thing and we can't do anything about the mining, we are at least going to make them do it the best way they possibly can to provide the least impact on wildlife.'

Small wonder he is not the most popular man in Wyoming and once had a mine engineer 'try to jump over my desk and beat on my body. Some people don't like me very much but that's the way it goes. I made the decision right at the beginning to establish myself as being relatively inflexible. I realized that if I compromised too much I was going to lose everything.'

In spite of the odds stacked against him, the approach seems to be working, possibly because the mining companies have recognized that Harry does represent an increasingly powerful ecological lobby. When he began the job he wrote an article which declared uncompromisingly: 'Wildlife managers and other biologists, with the exception of those working as environmental consultants, have generally viewed the great increase in mining activity in Wyoming as a major catastrophe.'

Wyoming laws now insist that mined land be restored, but hope for landscapes as badly scarred as this seems small.

Harry, as they say out West, was laying down the chips and for a long time he was a very lone voice in a very large wilderness. But recently there have been some signs that the wind is beginning to blow his way. Planning Department staff in Fremont County recently conducted a poll on urban and rural thinking on land use priorities. Three-quarters of the respondents to this survey agreed, or strongly agreed, that further urban development should be prohibited in areas where it would diminish numbers of big game animals. Also, and perhaps more significantly in American terms, most respondents were prepared to put their money where their mouths were. Protection of wildlife habitat was chosen as their first preference for the use of county and city funds, with protection of scenic areas as their second choice.

Perhaps reading this wind as well, a significant number of mining companies have started to take reclamation seriously and implement methods Harju's department has had ready and waiting for some years.

Topsoils must be treated like valuable minerals and carefully stacked to avoid sterilization. Replanting must aim at great diversity of vegetation, not simple grassing over. Harju insists that reclaimers take shrub cover very seriously and use seed mixtures containing several species. 'Reseeding only with grasses will replace food and cover for only a few species. A lot of this will not be available in heavy snow, whereas shrubs usually are. Shrubs also trap and hold snow which contributes to moisture conservation.' He also insists that the topography be carefully preserved. Slopes must be restored and rim rocks, which provide vital cover for animals, should not be buried under mining spoil.

It is all a very expensive process and irritating grit in the teeth of an industry that has never had to bother with much of it before.

It must also be admitted that while Harry Harju seems to be winning at home he is personally still pessimistic because of what he sees to be the national attitude to fuel. He thinks the drive for self-sufficiency is a false grail and that Wyoming will be raped for nothing because: 'This country apparently doesn't want to conserve anything . . . we simply don't *want to*!'

So perhaps, like Canute, Harry Harju cannot stop the waves of what amounts to an industrial revolution sweeping across Wyoming and drowning its prolific wildlife. There has never been a Shangri-la that has lasted. But Harry has determined to have a damned good try and in America today he has some powerful friends. He lets it be known that he will, and does, enlist the aid of vocal environmental groups and Federal agencies. He has even noted that employees of mining companies have begun voting in local elections against certain development plans proposed by their own companies.

'If you start getting into a place where a man hunts, fishes and camps and tear it up, he won't like that very well no matter who he works for. I see people all week and I get rather tired of them. I use my weekends to get away. I like to go and camp somewhere. I like to go and walk a mountain path and just sit there and let the wind blow off the aspens. Maybe an elk will walk through the meadow or there'll be some deer, and the birds are singing. I light a campfire and sit round it. . . . That seems to rejuvenate me. I can make it through the week fairly well after one of those weekends.'

We suspect that with Harry Harju at the wheel of his truck, Wyoming may make out for a little longer too. But what he is doing may be more significant than that.

Elsewhere in this book, naturalists of the eminence of Konrad Lorenz have warned that our glib exploitation of the planetary resource is approaching crisis point.

It may well be that the day of the gentle naturalist, the character with butterfly net and

Wild horses still run free across much of Wyoming but will they, and hundreds of other animals, be swept away by the industrial revolution now sweeping across this sparsely populated animal-Eden?

inner contentment on whom this book was posited, is a thing of the past. In his place, tough, meat-eating pragmatists with a modern awareness of the balance of man and animals may have evolved.

Human predation, which takes the form of well-managed and commercially orchestrated exploitation of nature, could not be controlled by the latter day Gilbert White's of this world.

The Harry Harju type, tough as his old cowboy boots and toting his Biology PhD like his ancestors once carried a Colt .45, could be our last chance.

ROY MACKAY

A Threatened Paradise

Baiyer River Sanctuary, Papua New Guinea

In the central highlands of Papua New Guinea darkness falls over the dense tropical rainforests of the Baiyer river valley. Away to the west, billowing banks of cumulostratus cloud, outlined against the darker silhouette of the rearing mountain range, forecast one of New Guinea's frequent downpours.

As the first heavy drops of rain begin to fall, Roy Mackay climbs the steps of his bungalow, crosses the kitchen where his wife Margaret is mixing baby food for some orphan animal and vanishes into the book-lined living room. Soon the strains of a Mozart concerto mingle with the thump of a diesel generator and the astounding cacophony of the surrounding forest. As the rain begins to drum on the tin roof, Mackay – almost shouting against the din – explains how he came to be running New Guinea's first wildlife sanctuary.

Long before he came to this island, from Australia, Roy Mackay was intrigued by the idea of living in a place so little affected by civilization: New Guinea came into contact with the modern world so very recently; virtually no Europeans settled on the island before the 1870s, no annexation by colonial powers took place until 1885, and only in the past 20 years or so has Western development made much impact on the wilderness.

Like so many other naturalists, it was the brilliant colours of the birds of paradise that first caught the eye of the young Roy Mackay. After World War II he was appointed to the post of Preparator of the Australian Museum in Sydney, and of all the exhibits he was particularly drawn to 'a diorama of birds of paradise displayed like jewels against a background of black velvet'. Since all but six of the

Roy Mackay: providing a sanctuary for wildlife.

With 12,000 species of flowering plants, including thousands of orchids, with 700 known species of birds and 180 mammals, the rainforests of Papua New Guinea must be among the richest expression of natural life on this earth.

42 species of birds of paradise are unique to New Guinea, Mackay knew that if he was ever to observe these incredible creatures in the wild he would have to find a way to join a field trip to that remote and undeveloped land.

His first chance came to nothing when a motor accident prevented him from taking a job as taxidermist to a New Guinea expedition. This disappointment was assuaged several years later when he was given the task of cataloguing a large anthropological collection in Port Moresby, New Guinea with the eventual aim of establishing a national museum.

On taking up his new post, Mackay discovered that there was no national collection of natural history specimens. Although the country had been visited by numerous zoological and botanical expeditions and although vast amounts of material had been collected, most of it had been shipped off to museums around the world. He therefore determined not only to add to the anthropological material, but to lay the foundations of a natural history section – by collecting the specimens himself.

With this aim in mind, Roy Mackay travelled for more than 10 years throughout Papua New Guinea visiting some parts of the wilderness that had scarcely, if ever, been penetrated by Europeans. In a land where there are no railways and few roads he explored in the only way he could, by air and on foot: 'Each time I fly I see parts of the landscape I would love to cover with a foot patrol, and make a mental note in case I manage to get there later.'

Papua New Guinea lies just south of the equator, on the eastern half of the world's second largest island (the other half is Irian Jaya), straddling northern Australia which at its closest point is only 100 miles away across the Torres strait to the Cape York peninsula.

Anyone who has visited Papua New Guinea or has tried walking through tropical rainforest will know that Roy Mackay's wanderings have taken him through some most formidable terrain. The dominant physical features of the landscape are the mountains: a switchback of razor-back ridges marching from the coast up to the central highlands where the peaks rise to more than 13,000 feet above sea level. To ascend or descend these mountains you must be prepared to hack your way through forest so dense that, according to Mackay, 'sometimes you can only get through a quarter of a mile in a whole day, the bamboo is so matted and so thick right down to ground level'. The vigour of the vegetation is determined by the climate, which in turn is a direct result of the mountainous nature of the country. Warm winds heavy with moisture from the Pacific ocean are forced up by the mountains into the colder regions of the atmosphere where they condense and fall as rain. According to Roy Mackay, in some parts of the island an annual rainfall of 280 inches has been recorded and 60 inches is an average. Hence the luxuriant vegetation, the blanket of green that covers virtually the whole of New Guinea and which is one of a very few large tracts of unspoiled and relatively unexploited tropical rainforest in the world.

But for how much longer it will remain so is a question that haunts Roy Mackay. Aware of what is happening in other parts of the world, he sees the economic and social pressures developing all around him; which is why, after leaving his museum job, he has devoted his time to the preservation of at least one stretch of rainforest where the unique flora and fauna of New Guinea can remain un-

Much of Papua New Guinea is accessible only by air or on foot, and even then it is tough going. Roy Mackay has travelled the country both ways.

disturbed by hunters, loggers and farmers. The little bungalow where the Mozart plays stands on the edge of this reserve, the Baiyer River Sanctuary.

As soon as the rain has stopped Mackay goes out onto the balcony where his wife Margaret is about to feed a young echidna. This little 'anteater' is a striking example of the rich and unique fauna of the rainforest: altogether Papua New Guinea has 700 bird species, 180 mammals (of which one third are marsupials), 70 bats, 56 rodents and two monotremes (egg-laying mammals) of which 'Willow' is a charming example.

Willow is a long-beaked echidna (*Zaglossus bruijni*). He is about three-quarters grown (when mature he will be the size of a small dachshund), and as he appears at Margaret

Despite the shortage of ants, Willow the orphaned echidna flourishes on a substitute diet.

Mackay's heels snuffling along after his supper, he reveals himself as a very odd animal indeed. Like the duck-billed platypus, the echidna lays eggs and suckles the young in its pouch. Its short, powerful legs and sharp claws are well designed for digging, while the robust body is covered in both fur and spines, the latter being more abundant towards the rear. At the front end is a long, pointed beak of a similar design and purpose as the South American anteater's snout.

When Willow is presented with his two supper dishes he chooses as hors d'oeuvres a bowl of earthworms, and as he sets about it the utility of the beak becomes apparent: from its lower side extends to astonishing lengths a sticky, grooved tongue which is used to draw in the worms, dirt and all. In the wild the tongue could obviously be employed to pull the worms from their holes and also to plunder ants' nests and termite mounds. When he has demolished the starter, Willow sets about a second dish, this time a disgusting looking gruel. Mackay explains that echidnas are difficult to raise and this mixture of baby food, minced beef, calcium, vitamins and egg-yolk is the result of many anxious dietary experiments.

Mackay emphasizes that 'Willow' is not just a pet name: it is the actual name given to echidnas in the area where this animal was found. In New Guinea there are over 700 different dialects and languages spoken by the numerous tribes, which means that each tree and flower and animal could have scores, even hundreds, of different names. Or it could have no name at all; unless it is edible or useful in some way, it is unlikely to have had a handle attached to it. Unfortunately for its future, the echidna has many local names, which suggest that it is not only edible but that it is considered a delicacy. 'It's certainly under pressure because it is very well liked as food by the native people. Anywhere within a couple of miles or so of a village echidnas would be wiped out; and where the forest is destroyed for timber or for gardens, there's no canopy to keep them cool and no leaf litter to support their food supply,' Mackay reports.

The echidna is never found in the coastal lowlands below about 4,000 feet or in the cloud forest over 9,000 feet; its territory is in the so-called mid-montane between the two, but even then it is not easy to observe. A nocturnal animal it sleeps during the day under a rock or a log or deep in a pile of leaf litter. Even a naturalist as widely travelled as Roy Mackay has only once spotted one of these strange and attractive animals in the wild – on the slopes of Mount Albert Edward. When he tried to pick it up he was amazed at its weight and strength, but because it allows people to approach it is still virtually defenceless and the easiest possible prey of the hunters from the villages.

Roy Mackay spends much of his time patrolling his forest sanctuary on foot or on horseback. Although after 15 years he feels totally at home in this dark, damp, slightly menacing environment – 'I'm a bit of a loner, I like the quietness of it' – he does understand how it may seem to others: 'A lot of people, I know, have premonitions of lurking snakes hanging down from the trees to grab them and all that kind of thing; but really the rainforest is a very safe place.'

As he makes his way through the green gloom he points out some of the minor hazards of the rainforest. An innocent looking broad-leaved plant with purplish undersides is carefully avoided: 'It's one of a number of stinging trees – like a vicious nettle – that brings you out in a painful rash if you brush against it. And then there are leeches . . .' Mackay recalls a certain area he visited during his travels that has a particularly bad reputation for leeches. He had forgotten the warnings until he noticed that the forest floor and the plants were crawling with them. When he paused on the path he could see dozens hurrying in his direction while others waited poised in the foliage, ready to ambush him as he passed. Since he was wearing socks and shorts he was defenceless and later found his boots were full of leeches and blood.

One senses that even Roy Mackay's all-encompassing love of the rainforest does not extend quite as far as the leeches – or the bush mites that cause intense irritation if they manage to reach the bare skin. They also happen to carry scrub typhus. But if these less attractive aspects of the rainforest help to keep people out of some areas, then Mackay would probably consider it no bad thing.

Tropical rainforest has been described as the richest expression of natural life on this earth and every few steps through the green gloom reveal something new. Roy Mackay emphasizes that in the sanctuary, the possibility of coming across a previously undescribed species is very real indeed. He has discovered several, including a small, grey-green freshwater crab.

If there is any rainforest left to study, it seems that there is enough work in New Guinea for generations of naturalists. Only half the island's fish are thought to have been listed; new reptiles and amphibians are always turning up and when it comes to insects, the job has hardly been started.

Pushing his way through thick undergrowth beside a slow, brown river Mackay discovers a green tree python slumbering on an overhanging branch. Apparently as it grows, this handsome snake changes its colour: when very young it is either bright lemon or canary yellow, or sometimes brick red, but when it reaches about 18 inches in length it casts its skin and appears in a bright acid green livery. This particular specimen is 3 to 4 feet long but Mackay has seen them reach 6 feet. 'They're usually quite

Much of Roy Mackay's time is spent patrolling the sanctuary on foot or on horseback. By far the most destructive predator he faces is man.

As a juvenile this green tree python wore a skin of speckled yellow. After several months this is replaced by a mantle of brilliant green.

inoffensive, but if you grab them and make very fast movements, they snap. They move around mostly at night, feeding on mice and lizards and sleeping birds. Another interesting snake found in New Guinea is Boelen's python. It is probably the only species of python in the world that lives as high as 10,000 feet.'

Of all the creatures in the Baiyer River sanctuary, the birds are without doubt the most spectacular. Papua New Guinea is still a bird watcher's Shangri-la, and those who are prepared to spend the time walking through the very difficult country can still expect to spot at least some of the birds of paradise – albeit only as a flash of flamboyant plumage glimpsed against the darkness of the forest canopy – and to enjoy the gymnastics, squabbles and flirting of the many species of colourful little parrots.

Roy Mackay has observed most of the 40 species of parrots native to the country, including the world's smallest, the lilliputian pygmy parrot. This comic little bird is less than 4 inches long and makes its home in holes bored in the nests of tree termites. Initially drawn to the rainforest by the magical beauty of the birds of paradise, he has now seen and photographed many of the 42 species that inhabit the islands, including the extraordinary Emperor of Germany's bird of paradise, which concludes its elaborate courtship ritual by hanging upside-down from a branch while unfurling and gyrating its silver and gold tail feathers like a can-can dancer's backside!

In his beautiful book on New Guinea, Mackay describes his encounter with one of the rarest birds of paradise, Wahne's six-wired parotia 'so-named because of three long, wire-like plumes, with oval-shaped flat feather tips, growing from above each of its ears. It is a spectacular bird – velvet black except for a breast shield of metallic feathers which reflect rainbow colours, a tuft of gold-tipped feathers on its forehead and a patch of iridescent blue on its nape. The parotia displays on the ground, dipping its head onto its chest and shaking its wire plumes about so much that is is a wonder they are not broken.'

Roy Mackay fears that time may be running out for those who wish to observe such marvels in the wild. He points out that most birds of paradise live in the coastal and mid-montane forest, which is the habitat most endangered by logging and human disturbance. 'The blue bird of paradise is the most severely threatened. It occurs only in a very narrow altitudinal band which happens to be precisely the level where there are most people destroying most forest.'

The Emperor of Germany's bird of paradise. Once top fashion houses paid huge sums for its plumage.

However, a more immediate threat is presented by man's continuing desire to adorn himself in the brilliant plumage of the male bird of paradise. It was these flamboyant feathers that first drew the attention of Western man to the islands now known as New Guinea.

Four hundred years ago, the explorer Magellan returned to the Spanish court with the skins of several strange birds. It was assumed that they must have come from paradise not only because of the magnificence of the plumage, but because the carcases had no legs. The way this conclusion was reached sheds some light on man's changing attitude

The Great bird of paradise, one of New Guinea's 42 species of the family Paradisaeidae.

On most afternoons male Lesser birds of paradise can be seen displaying on well-established 'display trees'.

to the natural world. Today our first (and correct) assumption would be that the legs had been removed; but the 16th century 'naturalist' reasoned that the bird was legless, that it was engendered and spent its life in the air and that it must therefore inhabit some ethereal paradise. As it was clearly in the interests of the purveyors of these highly prized birds that the anatomical truth should be obscured for as long as possible, it was the custom to mutilate the carcases before they were shipped to Europe.

In fact, the first comprehensive and scientific study of birds of paradise was not made until 1898, when R. B. Sharpe, an English ornithologist, described and illustrated 36 species. As only a further six have since been identified, it is clear that by the 19th century, interest in birds of paradise was intense.

Unfortunately most of this interest was commercial rather than scientific: since the *haute couture* houses of Europe and North America were prepared to pay large sums for the brightest and rarest plumes, thousands of birds of paradise perished to satisfy the demands of fashionable women. Obtaining these skins could be a hazardous business, and some traders were satisfied to purchase what they could from coastal tribesmen; bolder spirits ventured inland in search of rarer specimens and some failed to return. Undoubtedly, the unspeakable in search of the uneatable were themselves consumed by anthropophagous natives! Yet so rich were the rewards that the trafficking persisted even after World War I, when the new Australian administration had outlawed the sale of birds of paradise in the territories. There is at least one wealthy Australian family whose fortunes were founded by a grandfather who smuggled that precious plumage out of New Guinea and disposed of it to unscrupulous collectors. Fortunately the international demand has now declined and what remains of it is strictly controlled; but the threat to the birds is as great as ever.

In the name of progress, the native people of Papua New Guinea have abandoned many of their traditions and customs, but *not* their passionate attachment to their elaborate ceremonial dress. The traditional function of the feathers and warpaint was to intimidate the enemy, and today the men stubbornly refuse to abandon this customary attire because intertribal rivalry and even warfare is still very much a part of their lives. As Roy Mackay was able to show us, just one ceremonial outfit can involve the most appalling slaughter.

Two local tribesmen on their way to a celebration, or *singsing*, called at the sanctuary and allowed Mackay to photograph and analyse the materials that made up their head-dresses. The first one contained the plumes from five male Princess Stephanie birds of paradise, one superb bird of paradise, two lesser birds of paradise, the relatively rare vulturine parrot, the long-tailed buzzard, the goura pigeon and the male and female eclectus parrot, the two sexes of which have gaudy and quite dissimilar plumage. The second head-dress was slightly less elaborate but between the two of them they accounted for at least a score of rare birds.

Far from dying out, Mackay is convinced that this kind of personal adornment is becoming more elaborate: '15 years ago when I used to see the *singsings*, the revellers were certainly well decorated – but nothing like today. If you were to see some of the large bonnets worn by one of the coastal tribes you could find the feathers of as many as 30 male raggianna [Count Raggi's] birds of paradise in one head-dress; and recently I saw one man wearing 30 of the streamer feathers from the King of Saxony bird of paradise. As those birds only have two of those pennants on their

Count Raggi's bird of paradise, threatened by the destruction of its habitat and by human vanity.

Tribal head-dresses are becoming more elaborate. This one cost the lives of a dozen birds.

heads, this one man's vanity cost the lives of 15 birds. That's quite a slaughter.'

Mackay suggests that this increasingly extravagant dressing-up has something to do with the new mobility of the people. At a time when a tribesman never went further than the next valley, and then only to raid a neighbouring village, the materials for his head-dress and other decoration would be supplied entirely from local birds and animals. Now that roads – and more importantly aeroplanes – have penetrated even the most remote parts of the country, an extensive trade in decorative exotica has grown up.

Roy Mackay points out that one of his visitors is wearing the fur of the male spotted cuscus. This particular marsupial mammal is not found here in the Highlands and in former times would never have been worn by the local people. 'Now they're doing a lot more trading with the coastal areas where this particular animal comes from and that pretty chocolate and white fur has become very fashionable up here. Everyone seems to want to wear it.'

Unlike its spotted cousin, the ground cuscus is well established in the Baiyer River sanctuary. About the size of the domestic cat, this neat black animal is one of about 180 mammals found in New Guinea and, like one third of that number, it is a marsupial. Mackay says 'about' 180 mammals 'because most of them are nocturnal and there could be others to discover'. Although he concedes that such a discovery is unlikely.

Like the other cuscuses, the ground cuscus is really arboreal, 'it has earned its name because it does come down to the ground more than other cuscuses, and even nests on the ground occasionally'. Its naked prehensile tail is of considerable interest: it is extremely strong and is used to support the animal's weight as it climbs.

Although the cuscus is an agile climber, it easily falls prey to its principal predator: man. 'Around here' says Mackay, 'they're considered quite a delicacy. When the locals go hunting they go specifically for cuscuses – even more so than for the tree kangaroo.'

Now to those of us who are unfamiliar with the animal, the idea of a tree-dwelling kangaroo is hard to accept. Kangaroos were obviously not designed to climb trees. Evolution intended them to use their powerful hind legs to leap over the plains just as their relatives do in Australia, and not to scramble about in the branches feeding on fruit, leaves and even small birds. And yet, New Guinea has five species of these shy and little-studied animals.

Tree kangaroos are largely nocturnal and Mackay recalls that as he was walking through the forest one night, he was struck by a missile that turned out to be a half-eaten fruit. When he shone his torch into the tree tops, there was a pair of tree kangaroos chomping away at the fruit, nonchalantly discarding anything unpalatable.

Despite appearances tree kangaroos can move very smartly from tree to tree and Mackay confirms that they can jump quite remarkable distances. 'It depends on the type of forest they're in. In these very high mountains where you've got very big trees providing large "platforms" for them to jump onto, they can leap 10 to 15 feet, with the tail hanging out like a great flailing rudder. If the distance between the trees is much more than that, they will come down to the ground and hop across to the next tree. . . .

'We don't know much about them in the wild. There is still a lot of work to be done on tree kangaroos. I've had some questions asked by overseas zoos and I just can't give them the answers; not just because *I* don't know, but nobody else knows either.'

Mackay claims that what is true of tree kangaroos is equally true of New Guinea's other animals. 'Of course people have been here shooting and netting and taking skins for museum collections but as for actual field study, very few people have seen anything of their behaviour in the wild.' Mackay concedes that the obstacles facing anyone undertaking such studies are daunting. Many of the animals are highly specific, for example Matschie's tree kangaroo is found only in the mid-montane forest in the remote Huon peninsula on the north side of New Guinea. Mackay has walked through it and believes it is some of the wildest and steepest mountain country on earth. In the same area are three species of bird of paradise, unique just to that patch of rainforest 'and when that goes, so do all the birds'.

Tree-dwelling kangaroo: one of five species about which very little is known.

Unfortunately, although the high cost of mounting expeditions to such far-flung places often acts as a deterrent to natural scientists it does not inhibit the activities of mineral prospectors and lumber companies, and Mackay is fearful they may get there first. The intensive search for oil and natural gas in recent years has increased the risk.

In one respect, at least, Roy Mackay is like all the naturalists we talked to: he is a man gazing with wonder at the beauty and the colour and the design of the picture before him yet constantly glancing over his shoulder, fearful of the vandal with the axe and the madman with the knife.

To outsiders his fears may seem hard to justify: after all, it is well known that Papua New Guinea is the world's last unknown; that only 50 years ago much of it was unexplored, unmapped, unsurveyed, and inhabited by a savage, stone-age people for whom warfare was a way of life and cannibalism, in some tribes, a customary practice. All this is true; and yet you need only step outside the Baiyer River sanctuary to understand the cause of Roy Mackay's concern.

In the 10 years since the roads came, progress – or at least change – has been rapid. On the sides of the roads the forest has been cut down and burned to make new gardens. A familiar site is a group of bare-chested women completing the 'slash and burn' operation that will give them another couple of acres of cultivable land. It is a dismal scene with charred tree stumps and smouldering undergrowth and clouds of smoke billowing across the valley. The cleared land will soon be planted with coffee, and it is the development of these cash crops that poses such a grave new threat to the rainforest and its wildlife. Until the roads came the people of these parts were subsistence farmers: they grew enough to eat and that was the end of their labours. Now, throughout the Highlands, they grow coffee and cardamon seeds and raise cattle, and as this new economic activity demands more land, that means there will be less forest.

And that the major stimulus for this sudden demand for cash, in a society that very recently paid its debts in pigs and never handled money at all, should be beer is almost unbelievable. Until they were united as an independent nation in 1975, the territories of Papua and New Guinea had been administered by Australia for 50 years. During that time, prohibition was enforced, as it was the paternalistic view of the white administration that 'liquor is not good for the natives'. Unfashionable as it may seem, there are many who now think that the Australians may have been right. Since prohibition ended the sale and distribution of beer have become the nation's leading commercial activities and also a major cause of social problems. That it has also stimulated the demand for cash and hence encouraged the destruction of the rain forest is one of the more unexpected evils attributed to the demon drink.

Mackay is able to demonstrate all too clearly what is happening to his valley and it is the shades of green that tell the story. At the crown of the mountain ridge is the dense, dark green of the primary rainforest; as the eye moves down the slope it soon detects the bright viridescent patch where regrowth has just started on the site of some old gardens; further down is the patterned green of the new plantations; and stretching away into the distance are the thousands of acres of even greenness which have lost all their cover and form part of a large commercial cattle ranch.

Roy Mackay recognizes that the governments of developing countries are under huge economic pressures to harvest their natural resources, and that the most powerful of these is population pressure. 'This is most evident in the coastal areas. They're the easiest to reach for logging and they have the greatest

'Slash and burn' agriculture is the traditional way Papua New Guinea feeds itself. But now much more forest is cleared for coffee and other cash crops.

population. In the 15 years I've been here, I've seen a tremendous amount of cutting away. I used to do a regular weekend trip to a place about 50 miles inland from Port Moresby and I'd pass through only two little villages on the roadside. But now the squatters are in there, and up to three miles on each side of that road is cleared for gardens. That's population pressure.'

With such pressures exerted, Roy Mackay is worried that much of the rich variety of wildlife on Papua New Guinea, and its storehouse of knowledge, will be lost 'before we even know what we're losing'. He points out that in practical terms, the destruction of plants may be the most serious loss to mankind. 'Medicobotanists have hardly started the immense task of examining the medicinal properties of New Guinea's flora. Even the herbs traditionally used by the local people to treat their ills have still to be properly tested.'

It is a sobering thought that what is only just beginning to happen in Papua New Guinea is now an established and probably an irreversible trend elsewhere. Tropical rainforests are found in more than 60 countries within the tropical regions and altogether they cover a very extensive part of the earth's surface – in the region of three million square miles – and if there is one thing that naturalists like Roy Mackay are worried about the world over, it is the disappearance of these great tracts of vegetation.

Every year an area the size of England comes under the axe and under the plough and if the figures published by the United Nations Food and Agricultural Organization are correct, if the destruction continues at the present rate, there will be little or nothing left in 20 years.

The magnitude of such a catastrophe can easily be grasped by glancing at a list of animals that can survive in the wild *only* in the rainforest. Already, in equatorial Africa, gorillas, chimpanzees and leopards are under pressure, as are the jaguars, tapirs, ocelots, macaws, toucans and hummingbirds of South America. In South-East Asia, the plight of the tiger is well known, but what of the wild oxen that used to live in the forests of Cambodia and Vietnam and which may already be extinct? And what is to become of the orang-utans and gibbons and hornbills and the other familiar animals of that part of the world? In another 50 years shall we be able to see the superb birds of paradise of Papua New Guinea only in a zoo?

New Guinea has so much that is unique and beautiful and without compare – the world's *largest* pigeon, *smallest* parrot, *biggest* orchid – Roy Mackay is fighting a lonely battle to save what he can for New Guinea and for us all.

Reference

MACKAY, R., *New Guinea*, Time-Life Books, 1976.

The Baiyer River valley, first wildlife sanctuary.

NEVILLE COLEMAN

Getting High on that Sinking Feeling

Caringbah, Sydney, Australia

Neville Coleman wears a gold medal round his neck on which are inscribed the words 'when your comfortable your dead' (sic). Both the sentiment and the mispelling provide clues to the nature of this complicated and intriguing man.

'Actually I was a dunce at school', he confesses. 'My father wasn't very well educated so he couldn't teach me anything, and because of problems at home I didn't get on very well at school, with teachers or the kids; everything I know I've taught myself; I've had 17 years in the biggest university in the world: the ocean; and that's what I call formal training.'

Certainly any small difficulties he may have had with spelling have not prevented Neville Coleman from becoming one of Australia's leading underwater naturalists; but what does he really mean by 'when you're comfortable you're dead'? 'That's just a little philosophy, when you set out to be better than anybody else at something you have to remind yourself to keep up the pressure. If you start to behave as if there's nothing else in the world to discover, if you've finished seeking, you sit there and vegetate till you die.'

There seems to be very little chance that Neville Coleman will stay in one place long enough to take root, let alone to vegetate. He is bubbling, restless and hyperactive; he responds to questions with enthusiastic monologues filled with anecdotes, detailed observation and his own brand of homespun philosophy. You do not have to listen to him for very long before realizing that he is totally obsessed by one thing: a self-imposed objective that he knows very well may be quite beyond him, but which he pursues with

Neville Coleman in his element.

Like most of Australia's coastal waters, the reefs surrounding Lord Howe Island are immensely rich in marine life. It is here, in the coral zone, that Neville Coleman explores and photographs his underwater world.

Coleman's urchin shrimp, *Periclimenes colemani*. Coleman longed to discover one new species; so far he has found 150!

frantic devotion. Neville Coleman has decided that single-handed he will photograph the entire marine fauna of the Australian continental shelf. 'Ten years ago', he says, 'that was a joke, and may be it still is; even so, I've photographed 10,000 animals on my own without any financial support. During that time I've taken a four-year expedition 40,000 miles right round Australia's coastline just to get the basis of the knowledge that I need . . . now I can actually identify 6,000 species in the water and know what I'm looking at.'

Since returning from that expedition Coleman has identified and cross-referenced around 30,000 transparencies of marine animals and plants for what he calls the Australasian Marine Photographic Index. It is the first photographic subtidal survey ever attempted on an Australia-wide basis and already the largest scientifically designed visual identification system in the southern hemisphere. And the work continues; indeed according to Coleman, it has only just begun, for there could be 100,000 waiting to be photographed. His biggest ambition *was* to find *one* new species. So far he has discovered 150, and some nine or ten have been named after him because they were previously quite undescribed by science.

Neville Coleman is justly proud of what has happened to that 'ignorant kid' from the Sydney slums, that little Neville who first went down to the seashore to forage for shellfish and firewood for his improvident family and who now mounts major marine scientific expeditions. 'The ocean meant a lot to me in those days because it was a source of food; a lot of our meals came from subsistence fishing. All my spare time I was at it, collecting oysters off the beaches, smashing mussels for bait and rowing great 14-foot boats round Sydney Harbour and getting blisters and back-handers for complaining. . . . I was very, very keen on fishing and not once did I ever come home without a crab or a fish or something; I would never come back empty handed; and I really learnt a lot from those days.'

But it was not until much later that he started to think of the ocean as something more than a source of food. When he left school he wanted 'something different from the surroundings I grew up in; I always had a feeling for doing the unusual.' His first step was to take an apprenticeship in photo-

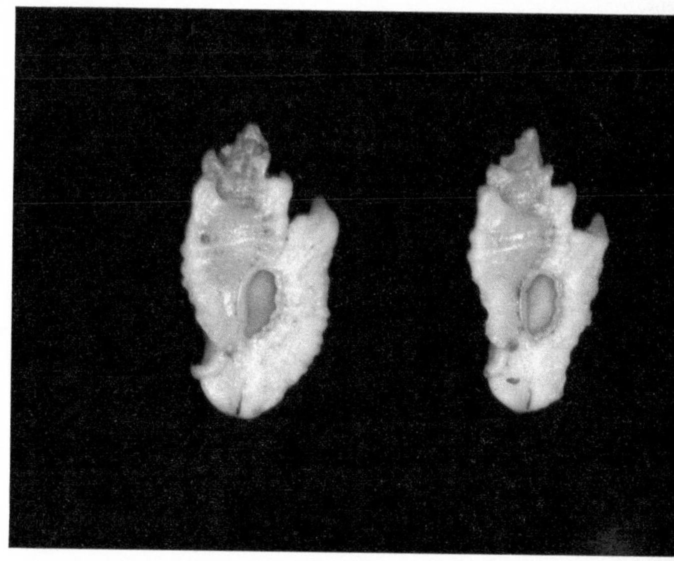

Above: *Pterotyphis lowei colemani.* Another new discovery that bears Coleman's name.

Below: This Stingaree was photographed off western Australia. Coleman travelled 40,000 miles to lay the foundations of his vast picture collection.

lithography: he did well in the trade and learned a great deal about photography and photographic reproduction, but still felt unsatisfied and unfulfilled. 'I had a pretty good job but I still had this urge to do something more than other people in the street; I wanted to do something which nobody else had ever done and I wanted to contribute something towards the search for truth.'

There is nothing pretentious about Coleman's struggle to describe the demon that drives him: 'I had this need to discover something for myself and the closest you can get to truth, the real nitty-gritty of things, is nature, because it doesn't lie; people may lie, but animals don't and that appealed to me.' And so in 1963 in his late teens, Neville Coleman began to spend his spare time diving in Sydney's Harbour and coastal waters. And the fact that he chose to study marine animals was an act of considerable courage: 'I was totally scared of the water, but then as a child I was scared of everything: my Mum, my Dad, the kids at school . . . and when I set out to beat my fear of the ocean, I set out to beat all those other fears too. When you get a beating as a kid for putting your hands out of the boat because "a shark will bite them off", you really get scared of going in the water; because you believe what the big people tell you. The first three weeks I spent under the ocean, I kept going in circles looking for the things that were going to eat me. But of course when I survived, I realized that all the danger is in people's heads, not in the water; and if you can conquer your fear, the whole world is there for you.'

For five years, underwater exploration was Coleman's spare-time activity, until in 1969 he threw up his job and set off on that four-year voyage of discovery around Australia's coastal waters. Since then he has written 10 books on marine life and his photographs have been published in books and magazines all over the world; and as far as he is concerned, his life of exploration has only just begun.

Earth's largest living space is provided by the oceans; they cover 71 per cent of our planet's surface, an area of 139 million square miles and in the deepest places, the ocean floor is more than 6 miles from the surface. This great body of water supports everything from the 150-ton blue whales to the simplest unicellular organisms. The areas that hold the richest marine life mostly lie close to land where the sea is shallow, shelving gradually from the tide line to about 500 feet and it is here, on the continental shelf, that Neville Coleman explores and photographs the teeming marine organisms.

Australian coastal waters are the most magnificent hunting grounds for the naturalist. Much of the coastline lies within the coral zone (between latitudes 30° south and 30° north) where the mean water temperature during the coldest part of the year never falls below 18°C. Apart from those comfortable diving temperatures, the water also has outstandingly good visibility, particularly on the coral reefs. Coleman has carried out more than 10,000 dives, many of them alone and many at night and he claims that very few of them fail to produce something that is new and exciting. We joined him for a few weeks when he had for once abandoned the shores of his native Australia and travelled north to visit a marine park established by a fellow diver, Bob Halstead, off the coast of Papua New Guinea.

There are three kinds of coral reef: coral atolls – a ring of islands that enclose a lagoon,

A photograph of a freshwater crayfish preserves the colour and form of the specimen as no other method can.

fringing reefs – found in shallow waters close to shore, and barrier reefs, like the Great Barrier Reef of Australia, which occur some distance off-shore and enclose a lagoon between the reef and the mainland.

It is a stretch of barrier reef just north of Port Moresby, on the south-east Coast of Papua New Guinea that Bob and Diana Halstead have developed as a marine park and nature reserve. Local villagers and recreational divers have co-operated with the Halsteads by not fishing the area, so that over a number of years, according to Bob, 'we've created a place where we have wild creatures that have become our friends; we know them personally, they know us personally. If you dive here you're immediately surrounded by many, many animals; they're not scared away; in fact, they'll swim right to you: big ones, small ones, incredible sizes and colours. What we're trying to do is to give people a beautiful experience in the ocean.'

The Halsteads knew Neville Coleman by reputation and they know that he shares their views about man's place in the ocean. 'I have Neville's books at home', says Bob 'and it's very valuable to have him here in New Guinea. As a naturalist he believes, as we do, that we have to change the idea that divers have had in the past: that when they go into the ocean they're "braving the depths"; what we're trying to do is to get people to be in harmony with the ocean, to be part of it and to have a completely magical experience.'

After 10,000 dives, still excited as a child.

Naturalists like the Halsteads and Neville Coleman believe that as long as divers continue to regard the ocean as a hostile environment, they will enter the water in a destructive and predatory frame of mind. Of course, in trying to change this attitude they are fighting a cultural rearguard action. There is such a huge body of myth, legend and fable about 'the denizens of the deep', that it would be surprising if we were not profoundly influenced by it. Undoubtedly, seamen of the past who sailed the wild and wasteful oceans in flimsy vessels are responsible for leaving us a dreadful inheritance of monsters: of 600 foot sea-serpents and sirens to sing us to our doom: of giant octopus and squid: of mermaids and deities and lost continents. And where the ancient mariners left off, the modern cinema has taken up the story.

As the Halsteads' boat carries us over the reef, Coleman prepares his diving equipment and his cameras. Despite the hot day and the warm water, he wears a complete wet-suit. He explains that this is to protect him, not from the cold, but from the coral: when diving every day, a tiny graze incurred in the water will not heal and may develop into an abscess – an anathema to Neville Coleman, as it could limit his diving.

Coleman is preparing three underwater cameras: 'with these I can photograph everything from plankton right through to whales; we could find anything down there and I've got to be prepared to get the pictures'. Considering he has made over 10,000 dives, by the time we reach the reef Coleman is in a surprisingly high pitch of excitement but nothing compared to his explosive state when he resurfaces after about 40 minutes down below: 'Honestly, that dive! There's more species down there than I've ever seen anywhere in my life. I got some fabulous pictures but not half as many as I'd like to get. If I dived twice a day for about four years I might get about half of what's down there – it's beautiful.'

Coleman is especially impressed by the coral formations. Reefs are built of calcium carbonate laid down by colonies of stony corals each of which has a hard outer skeleton, or coral cup. These tiny coral cups, which range from one to three millimetres in diameter, are continuously added to, growing into

The green-backed wrasse, just one of our 200 fish so far identified in the waters surrounding Lord Howe Island.

distinctive shapes that suggest their names: whip corals, sea fans, pipe corals, sea pens and sea pansies. Some corals are brightly coloured due to the presence of minute organisms that live symbiotically in the coral tissue. 'Imagine', says Coleman, 'beautiful coloured trees, four or five feet high, not one or two, but hundreds of them hanging to the cliff-face that drops straight down 60 or 70 feet. There must be at least 30 species of sea fans and they're the biggest I've ever seen.'

'If you went down there', he explains 'you wouldn't *see* the colours'. Coleman reminds us that light entering the sea is quickly scattered and absorbed, much of it dissipating in the top 3 feet. At a depth of only 30 feet or so most of the remaining energy is in the blue-green part of the spectrum. Since reddish objects will appear brown or black in this light, it is to be expected that many creatures have adopted camouflage coats of red pigment and that as Coleman illustrates with some of his specimens, is the case. He shows us a bright orange sea-cucumber. 'I've recorded about 150 species of these in Australia and I've never seen this one before. But of course on the ocean floor it looks brown. Sea-cucumbers are very useful they're like little vacuum cleaners. They move along the ocean floor feeding on detritus: it comes in one end and out the others. Usually they're just nondescript colours but this one is special.'

As the sea-cucumber is returned to the water, Coleman rummages in his collecting bag for a glass tube containing another small bright red creature. This time it is a nudibranch mollusc: a tiny sea slug with external gills. Coleman believes that nudibranchs are a very good example of the unique value of his photographic index. The traditional way of cataloguing marine creatures is to collect specimens and preserve them in glass jars, but according to Coleman: 'There's no method

Above: Only one animal underwater gives Coleman cause for alarm: a scuba diver armed with a spear.

Left: The shapes of corals often suggest their names: this is a staghorn coral.

known to science of maintaining their colour or their form; but in my photographic index I have 600 species of these [nudibranchs] all cross-referenced against preserved specimens; so now you don't need to collect the animal in order to identify it. We have an accurate record of the colour and the form and once you've seen the photograph you can just go into the water and use your eyes. They're beautiful aren't they?' He revolves the glass tube so that we can admire the brilliant colour and see the delicate filaments of the animal's external gills.

The nudibranch is a mollusc rather like a snail without a shell and although it certainly looks very vulnerable, according to Coleman it has evolved a very effective chemical defence: 'Special bad-tasting substances which they can squirt out when the fish grabs them; and the fish soon learn, they're pretty smart. Some juvenile fish might take a nudibranch but they'll spit him out again because he doesn't taste nice.' We assume that Neville has already tested this theory himself; of course he has, and he holds the same opinion as the fish. Nor does Neville think that such an experiment is in the least eccentric. 'It's the only way to learn anything about the ocean', he says, 'there's no one to teach us down there, except the animals and if we don't learn from them we shall soon be back in the dark ages'.

Coleman constantly returns to his favourite topic: man's ignorance of the ocean and worse than that, his arrogance: 'We judge everything by ourselves and yet we don't even understand ourselves. How then are we supposed to understand other creatures? The more we find out about the natural processes of life in all sorts of animals, how they run their systems and how they survive, the more we can learn about the chances of our own survival on this planet.' He turns back to the tiny sea slug in its glass tube. 'When you mention "slug", most people want to shy away or squash it, and yet you can see that they're incredibly beautiful, not aggressive: any diver with a little bit of understanding of the ocean is certain to fall immediately in love with them.'

During the dive, Bob Halstead had introduced Coleman to some of their 'tame' fish. One of these, a large moray eel, has made its home inside a wrecked fishing boat and regularly makes its appearance through one of the portholes. 'Underwater', says Coleman, 'it appears to be about two metres in length. It's certainly bigger than any that I've seen in Australian marine parks. It's a rather beautiful animal and is certainly not aggressive.' Those of us who have had the odd glimpse of the forbidding features of the moray, peering from its hole in the coral, may be inclined to disagree with Coleman about its instant appeal. And yet once again it seems that the animal's behaviour belies its appearance.

Hexabranchus flammulatus, the Spanish Dancer, is one of the small but colourful animals Coleman particularly admires.

Members of the family muraenidae, or moray eels, spend most of the daylight hours lurking in crevices in the rocks. When the creature is disturbed, the head appears and rows of long depressible canine teeth are visible as the jaws regularly open and close. It is this baring of the teeth that gives the moray its vicious expression, but according to Coleman 'when their jaws move like that, they're not threatening, only breathing'.

Morays are all carnivorous, emerging from holes at night to feed on fish, crustaceans and octopus. It may be that the techniques employed by the eel to kill and consume its prey have contributed to its ferocious reputation. Its teeth are designed to tear rather than to cut flesh, so once the food has been detected – using 'tastebuds' on the outside of the lower jaw – it grabs a mouthful and holds on. And then, according to Neville: 'it coils its body almost in a slip knot, forcing its head against its prey, and then violently uncoils and rips the piece of flesh straight out of the bigger animal . . . it's incredible to watch.'

Under normal circumstances moray eels will do their best to avoid divers, but the one that lives in the wreck has become 'tame' in the sense that it has become accustomed to human visitors and will emerge from its porthole to be fed and handled. Coleman says that compared with some tame fish morays are particularly gentle, that they never snatch the proffered food. 'I know cases where divers actually feed eels with fish from their own mouths. Until several months ago, I could honestly say that in all the years I've been diving, I have never been attacked by an eel, but then, all of a sudden, I was. And I still don't know why. It just shot out of its hole and bit me on the finger; luckily I was wearing a glove . . . there were a pair of them and I assume they were male and female, and he was telling me to get the hell out of it! I was the offender, not the eel.'

Also on the dive Coleman had come across several giant clams; which together with sharks and sea snakes and moray eels, are the favourite underwater villains of the entertainment world. They are supposed to drown divers by trapping their legs. When we visit a marine research institute run by the University of Papua New Guinea, where there are several giant bivalves kept in tanks, Coleman is able to demonstrate just how absurd the 'big clam stories' really are. Each clam weighs around 1000 lb, and it is possible to see how it got its evil reputation. When it is undisturbed the two sides of the clam shell remain open, but as soon as it is alarmed it closes rapidly and with sufficient force to send a spurt of water over the side of the tanks and

A fish at home in the coral reefs – the brightly coloured, crimson-banded wrasse.

The giant cuttle fish is quite harmless and yet, with 'giant' clams and 'giant' squid, it is part of the underwater demonology that Neville Coleman hopes to exorcise.

A black cowry set off to perfection by its bed of scarlet coral.

into our shoes. And there is no doubt that the adductor muscle which opens and closes the shell is extremely powerful: it has to be to move that weighty shell with any speed. Hence the story that it closes on divers' legs with sufficient ferocity to trap or even amputate them.

However, we are able to satisfy ourselves that such a catastrophe is quite impossible. Following Coleman's lead we plunged our arms up to the elbows, into the clam shells. The creature certainly closes up instantly; but the edges of the shell are so well upholstered with very soft, slimy tissue that withdrawing the arm is simple. It is a curious sensation but not in the least unpleasant or dangerous and according to Coleman 'much more likely to harm the clam than the diver. Some people poke them with a stick just to see them squirt water, but you see they hurt the animal and very often they die'.

Apart from this casual and senseless destruction, the giant clams have almost been wiped out on some reefs by the activities of Japanese and Taiwanese poachers. The edible part of the clam is the adductor muscle that closes the shells; the Chinese use it to enhance

Everything underwater appears much larger than life, which makes this giant crab a monster.

the flavour of other fish, and the canning industry is prepared to pay up to £30 per lb for it. Since one animal weighing over 1,000 lb yields only 6 or 7 lb of this meat, and because clams take 40 or 50 years to reach maturity, the scale of the destruction is appalling. The Great Barrier Reef of Australia has been particularly badly hit: 'It's just like a graveyard of big clams. The poachers swarm over the reefs, open the shells, cut out the adductor muscles and take them back to the mother ship where they're frozen and sold in Hong Kong for vast sums. They just leave empty clam shells over the entire reef – a terrible mess.'

At the research institute a team of marine biologists is investigating the possibilities of farming clams, or at least of restocking the reefs that have been plundered so indiscriminately.

Clams are very suitable for aquaculture because of their feeding habits. They simply filter suspended matter, plankton and organic detritus, from the surrounding ocean. The water is drawn onto the surface of mucus-covered gills which trap any food particles and pass them on to the mouth. Coleman explains that their feeding habits can, however, produce undesirable consequences for man. 'The danger is not to divers trapping their legs but to those who eat clams. By filtering suspended matter from the water these animals can concentrate dangerous coliform bacteria and hepatitis virus in their bodies.

What is more, when a certain organism, Goniaulax, occurs in the plankton, filter feeding clams may concentrate a powerful toxin which is dangerous to man and to fish; it's one of the agents responsible for the infamous 'red tides' that cause massive fish mortality along some coastlines. But the wonderful thing about clams is that even if you keep them in water that is completely clean and has no suspended food material in it, they will still grow by using their own 'farming' mechanism. On the surface of the mantle which is exposed when the shell is open, they harbour algae, little plant cells that respond to sunlight and make food that is passed on to the clam itself. As well as growing the giant clams in tanks, the researchers have successfully 'planted' colonies of small clams in cages out on the reef where they are now flourishing.

Coleman is attracted to another tank at the institute, which appears to be uninhabited but on closer examination contains a handsome snake partially obscured by a rock. It is one of a considerable range of sea snakes found in these waters and is another of the creatures that Coleman feels is particularly maligned and persecuted because it is considered dangerous. The banded sea snake, which now slides off the rock and begins to swim slowly round the tank, is certainly venomous; in fact its bite is much deadlier than that of most land snakes. This particular specimen is a large female about 5 feet in length with the broad, spatulate tail characteristic of water snakes. Unlike many other species, the banded sea snake spends part of its time on dry land: it comes ashore to lay its eggs during the breeding season, and also to digest its food and bask in the sun.

Coleman finds their behaviour particularly interesting because all the species he has come across around Australia stay in the water and bear live young. The same day, he had seen several examples of this species under water: 'I found them a bit difficult to photograph because they wanted to go to sleep which was disappointing as I would have liked to have observed them working and catching their food . . . they are potentially dangerous, like many things in the ocean – I'm potentially dangerous too. If I'm afraid, my fear is dangerous. Fear is not a good thing to enter the water with because you're going to make a mistake and set off a reaction in the animal that can make it aggressive. It's not the animal's fault, it's yours.

'I think people are the most offensive things in the ocean because of their ignorance of how other animals live.' Coleman claims that the only creature that now really alarms him underwater is man: 'especially when they have some sort of weapon which they're using to protect themselves. I've been in the water with divers armed with spearguns, who were supposed to be protecting *me* but who put me in far more danger than I would ever have felt from any of the animals. People are scared and if anything happens they'll pull the trigger; I've seen what those spears do to fish and I just don't want to be in front of them.'

Neville Coleman knows that only when he had conquered his own fear of the ocean could he begin to learn about it. 'If you go into the water as a predator or as an intruder and just

Despite *Jaws*, Neville Coleman hopes to convince us that we can share space with animals like the Galapagos shark.

blunder around the bottom you're going to learn nothing. The animals are there; if you want to know a little bit about them, you *can* share where they live totally for the time that you're down there. And remember, you're only there on *borrowed* time: you've only got that little bit of air on your back. That's why observations under water are so sparse in comparison with what other naturalists can do: they can spend 24 hours a day watching their animals; 30 or 40 feet underwater, we've got one or two hours and that's not very much time to find out how it all works.'

After 10,000 dives, Neville Coleman is doing his part in finding out how it all works, but he feels that his most important task is to educate others about the ocean. Eighty-seven per cent of Australia's population live within 20 miles of the coast and yet Coleman says, 'only very, very few, perhaps one per cent of us ever go down to the sea with a face-mask and a snorkel, because of the sharks and sea snakes and all the other creatures that we're ignorant of but which scare us.

'If only I can convince a few others that after you've been down there a while and you see how the animals live, you know their reactions and you understand them, you become part of the ocean, and then it starts revealing its secrets to you.'

Reference

COLEMAN, Neville, *Australian Marine Fishes in Colour*, Reed, 1974.

COLEMAN, Neville, *A Field Guide to Australian Marine Life*, Rigby, 1977.

COLEMAN, Neville, *A Look at the Wildlife of the Great Barrier Reef*, Bay Books, 1978.

COLEMAN, Neville, *Australian Beachcomber*, Collins (Australia), 1979.

DENSEY CLYNE

The Big Game in the Back Garden

Turramurra, Sydney, Australia

In the days when Australia still sought to attract migrants from Europe, the recruiting posters probably described it as a 'land of opportunity'; and of course it is. Not least, it provides the chance for the Australian people to live in an astonishing range of natural settings and climatic conditions: everything from rich farmlands to burning hot deserts, from terraced vineyards to coral islands with bone-white sands. And yet, despite this wealth of exotic surroundings, most Australians choose to settle in unremarkable suburban streets. More than half the population is concentrated in the two cities of Sydney and Melbourne and most of the rest live in the other provincial capitals.

If you take the train that crosses the Sydney Harbour Bridge to the northern suburbs, after 40 minutes of leisurely travel you arrive at Turramurra station. Turramurra is one of those delightful names – presumably derived from the aboriginal language – that resound of the outback: of ghost gums and kangaroos and an empty sun-baked landscape untouched by the enterprise of man; but Turramurra does not live up to such expectations. It is everything that from the writings of Dame Edna Everage we have learned to expect of the gladioli-strewn Australian suburbs.

This is the land of the timber bungalow and the white picket fence: of the neat garden planted with introduced shrubs and flowers to remind the settlers of the familiar landscape of some ancestral European home. It is not a likely place to find a naturalist who loves the wilderness and who has become one of Australia's best known natural history writers and photographers; and yet Catalpa Crescent, Turramurra, is the chosen home of Densey

Densey Clyne, revealer of fantastic invertebrate secrets.

Australia has more than 1,500 species of spider, many of which can be found in the ordinary suburban gardens.

Clyne and the key to her passion for the natural world.

Densey Clyne has discovered that even a suburban garden 'is a kind of nature reserve. No matter how tiny or how tidy, the wilderness is there, arbitrarily partitioned by fences that bear no relation to the territories of any but the human occupants.' It is this wilderness, the 'garden jungle' as she calls it, that Densey watches and photographs and studies, and which she has made the subject of a popular book and film. And a glance at a couple of her other book titles: *How to Keep Insects as Pets* and *A Guide to Australian Spiders* puts her real interest in focus.

Densey has an enthusiasm for the invertebrate animals: for the insects and spiders, slugs, snails, caterpillars and their kind which most of us fail to observe, and which, when we do spot, we are more likely to attack with a shudder and an insect spray than drop to our knees to admire. All too often our interest in the natural world is so fixed on those furry vertebrate star performers – the lions and tigers, the pandas and bears – that we fail to see that there are animals more spectacular than those, more colourful and amusing, more deadly and certainly more diverse, in every city backyard and suburban garden.

Not that Densey Clyne's Sydney garden is entirely typical of Turramurra. Where her neighbours' lawns are neatly defined by paving stones and geometrical flower beds, Densey's daisy-speckled grass recedes into a tangled mass of untended vegetation, shaded by the crowns of huge eucalypts. At first the neighbours were astounded to observe this attractive dark-haired woman on her hands and knees, day and night, in fair weather and foul, rummaging in the shrubbery. Now, like so many naturalists, Densey is simply regarded as a harmless eccentric who can even be consulted about sick birds and unusual moths and caterpillars.

For many of us, nosing around other people's gardens is a joy we are even prepared to pay for. If the tour is conducted by the proud owner then the experience can sometimes be seriously diminished, but not in the case of Densey Clyne. Her 'garden jungle' lies close to a main highway and a railway line but the sound of the traffic is drowned out by the cacophany of bird cries. To call it bird *song* would be misleading since many of the species attracted to Densey's birdtable and swinging trays of food are not the possessors of nature's greatest voices. We discovered her feeding the rainbow lorikeets with a soggy mess of bread and honey, the brilliant little parrots screaming and scolding as they swooped down from the tall eucalypts. Some of the birds are so tame that they will take sugar from Densey's hand. She believes there are about 47 species of birds in her garden 'and seven of those are parrots. It's quite incredible, parrots flying around freely in your garden like this'.

Densey's wonder at the tropical birds in her garden and her very slight Australian accent betray her English parentage and upbringing. She came to Australia at the age of 12 and spent most of her childhood 'thinking of England as a lost paradise'. She had an idealized view of nature based on books illustrated by artists like Arthur Rackham, and for some years she was unable to see and appreciate the marvels of the Australian countryside; it was far too unlike that distant English fairyland. But today she deplores the actions of the Aclimatization Societies in the last century, who introduced numerous creatures into Australia. 'I know they were homesick and they wanted familiar birds and

Australia's teeming bird life numbers 745 species. Rainbow lorikeets are among seven species of parrot found in Densey's garden.

animals around them. So in came the bunny rabbits and the sparrows and the Indian minahs and I heartily wish I could get rid of some of them. I do like all kinds of birds but it's a pity to see the nesting sites of native birds taken over by sparrows and starlings; they get totally out of hand because they don't have natural enemies to keep them under control.'

Knowing that Densey's main interest is insects, the birdtables came as a surprise but she says that there is not too much conflict as only a few of them are insect-feeding birds, most, like the rainbow lorikeets, being nectar feeders.

Densey explains that their main source of food is the eucalypt: in each of the big gum trees that surround the garden, gallons of nectar is produced. In the flowering season the trees froth out in creamy blossoms, each of which is a little cup of nectar surrounded by masses of pollen. 'For the lorikeets it's the pollen that's the main source of food; they have what's called a brush tongue – a furry tongue, with which they lap up the pollen from the ends of the flowers. They also drink the nectar, and follow the blossoming of the different species of gum trees. They're totally nomadic. That's why there are sometimes a lot around here and sometimes only a few. Three or four never go away at all; I think because they've become dependent on my bread and honey.'

Another engaging bird to visit Densey's garden is the kookaburra, the largest of the Kingfisher family. She stands at the kitchen window calling him, and before long the ungainly bird with its big head and watchful eyes perches on the branch a few feet from her outstretched hand waiting to be fed with

The kookaburra has invaded Australia's suburban gardens and is readily hand fed with cheese.

lumps of cheese. Although a foreign food, it seems that many birds like the taste of cheese: 'I think it must have the same kind of texture as beetle grubs; perhaps they regard it as a kind of processed grub – I don't know, I've never tasted a beetle grub.' That final admission almost comes as a surprise: Densey Clyne has such an intense curiosity about the creatures which share her house and garden, that sampling their diet would be wholly unexceptional behaviour.

As you wander through her garden jungle you are frequently challenged by her to attempt what she has learned to do so supremely well: 'to become increasingly aware of a shadow world, existing alongside our own, whose miniature inhabitants I see as the real possessors of our gardens.' She has learned to see what most of us will miss because we are not really looking and because so many insects are masters of camouflage. She points to the overhanging branch of a gum tree. At first sight there seems to be nothing of special interest, and then it becomes clear that a small bundle of twigs, about 4 inches long, is on the move. It is a superb example of what Densey calls 'the craftsman caterpillar'. The casemoths (Psychidae) are the master builders of mobile homes and none does a better job than the Saunder's casemoth (*Oiketicus elongatus*). This is the largest Australian species and, according to Densey, 'it's common in gardens, feeding on several shrubs, but I usually keep one or two indoors on a branch of eucalypt or angophora'. Behind that modest statement lies a beautiful and detailed piece of research work.

For more than a year, Densey Clyne observed, filmed and photographed four Saunder's casemoth caterpillars and recorded the construction of their ingenious portable shelters. Like other casemoths the Saunders caterpillar begins by weaving a silken tube – 'that silk is extremely tough; if you want to get inside and see what's going on in there, you have to cut it with very sharp scissors. It's entirely a matter of protection and camouflage'. The camouflage, in the case of the Saunders casemoth, is a pattern of short sticks attached to the case and when Densey observed in detail how the task was performed by the caterpillar, she found it to be astonishingly complex.

First the caterpillar attaches itself securely to a suitable twig with a silken safety belt; then it gnaws away at the stem just ahead of the attachment until the leafy tip drops to the ground. Next it releases the safety belt, backs down the stem a little and refastens it. Here, a second cut is made which frees a 3 to 4 millimetre piece of stick. 'It never does that silly human thing you see in cartoons of sitting

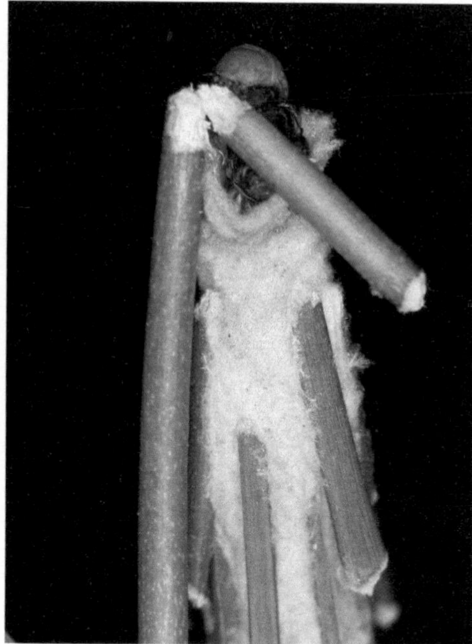

The silken bag of the Saunders casemoth larva can be up to 6 inches long. It is festooned with short sticks cut from the food plant by the caterpillar and attached with silk at random all over its surface. The sticks are of uniform length and thickness and are added, several at a time as the case is enlarged, by weaving more silk around the neck.

on the wrong end of the branch it's going to saw off . . . it never faces the wrong way.' The caterpillar then fixes its stick to the neck of the case and vanishes inside. 'All you can see for a while is a lot of humping and bumping as it moves about and then you can see that it's cut a hole in the silk where it's going to put the stick.' Next the caterpillar emerges through the new slit, detaches the stick from the mouth of the case and places it above the newly made opening. Finally, while the stick is held in place by the three pairs of thoracic legs, the caterpillar weaves it into the wall of its home and closes the slit.

Like other grasshoppers, this pink-faced Tettigonid is admired for its lugubrious expression and engaging habits.

This extraordinary sequence of tasks is just a part of a long life cycle that demonstrates successive patterns of elaborate reflex behaviour, and it is hard not to agree with Densey that, in comparison, the study of vertebrates is dull and predictable.

Of all animals, insects show the greatest diversity: they can claim more species than the rest of the animal groups put together. And how little we really know of them! As we continue our walk around Densey's garden, a line from one of her books comes to mind. 'I am a Gulliver, an awkward giant whose every step can endanger lives, because there is not one square foot of vegetation or tree-trunk, or bare earth – or even fence – that doesn't have its small and vulnerable occupant.' And for the naturalist, the exciting corollary to that thought is the fact that many of these 'small occupants' are still unknown to science, undescribed and unstudied.

For the ornithologist or the mammologist the chance of discovering a new animal is remote. Their melancholy task is far more likely to involve the chronicling of vanishing species than the detection of new ones. And yet, as we pass her small, muddy garden pond, Densey Clyne is able to say, quite casually: 'Somewhere in these leaves, there's a small grasshopper sitting that's never been scientifically described; it hasn't even got a name. I don't suppose it cares a bit. I've been living in this garden for about 20 years and I'm still finding new things; I've been watching this little grasshopper carry on its life history year after year since I've been here and found out most things that there are to find out about it.' The remark was made as a simple statement of fact, quite without hubris, and yet Densey responds warmly to the idea that the grasshopper might one day be named after her.

Although she has already had two papers published in learned biological journals, Densey has had no formal scientific training of any kind; in fact she dropped out of school at the age of 13 and she has no regrets. She feels that she might even be limited as a writer by formal biological training. 'I'm not a scientific kind of writer: I like to use a certain amount of imagination and be fairly subjective about what I write. I like to talk about my *feelings* about animals and I think that most biologists regard that as being not quite right.' It is obvious that Densey's response to the natural world is primarily aesthetic and emotional.

As we near the end of our garden tour she spots a large and striking lacewing and is carried away by the beauty of it. 'Look at those dark sapphire blue eyes . . . I call it the blue-eyed lacewing, which is a lot easier to remember than its latin name *Nymphes myrmeleonides*.' It is certainly a handsome insect with its black and yellow body, finely veined, transparent wings tipped with black and orange and those extraordinary eyes of the deepest blue. Densey explains that like other lacewings the female of this species mounts her eggs, as she lays them, at the end of a fine stalk stiff enough to hold the eggs erect. 'It's really fantastic to see, but not many people have ever watched it because, like most interesting things in nature, it happens at night. When other people are gazing at their television sets, I'm out here in the garden, either being eaten alive by mosquitoes or freezing to death in winter. But there's so much more drama going on out here: much more sex and violence than you've ever seen on television. Just think of the private lives of spiders . . . I can certainly tell you some tall tales about spiders.'

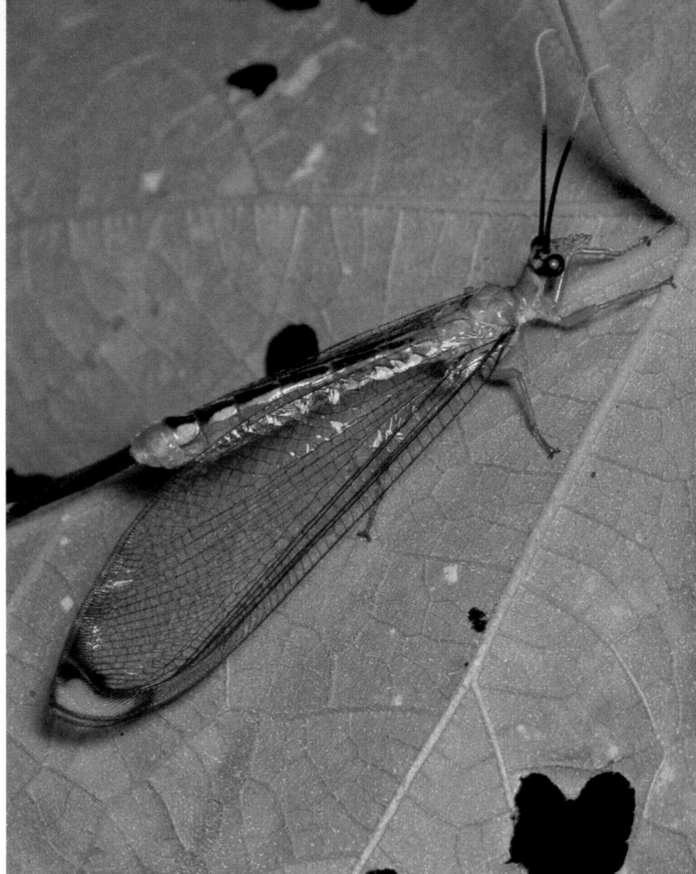

Nymphes myrmeleonides is more easily identified and remembered as the blue-eyed lacewing.

Many lacewings lay their eggs at the end of stiff, silken stalks.

Some of these tall tales Densey has vividly recounted in her award-winning book *The Garden Jungle*, and most of them concern the sex life of the spider, which must count among the strangest in the animal kingdom. Not least of these is the intrepid endeavour of the male St Andrew's cross spider to court and seduce a mate.

'The female of the species is said to be more deadly than the male: in the case of the St Andrew's cross spider this is all too horribly true. The male that courts this female courts death; he risks life and limb, and speaking of limbs, it is fortunate for him indeed that he starts off with plenty of them. . . .

'A mature female sits quietly in the sunshine at the centre of her web. On the web with her is a tiny male – actually sitting on the web itself, within a centimetre of the female. . . . He is not in that position for the purpose of mating, because that takes place on the outskirts. He is there for courtship only, and this takes a strange and dangerous form. It seems to be an essential part of foreplay among these spiders that while the female sits quietly at the centre of the web, the male should run around her tickling her legs with his feet. As she holds her eight legs in four pairs, the male makes four approaches every time he runs around her, tweaking and plucking, never staying still for an instant.

'The male also spends a lot of time . . . on the outskirts of the web, and from here he lays down a mating thread that runs to the centre of the web. His objective now is to lure the female on to it. To do this, he hangs upside-down on the thread a few centimetres away from her, plucking the thread vigorously, producing sweet spider music that she hears with her feet and finally responds to with her whole body. Slowly she turns to face him, and moves out along the mating thread. There is a brief moment of leg play as they touch, and now he has literally hurled himself on to this giant's abdomen and she is hugging him against her with all her legs. Yet all seems to be well; the male's palps are in place, and the female has not attempted to molest him. But her quiescence is only apparent, because almost imperceptibly, behind the male's back, something sinister appears to be happening. A broad ribbon of silk is appearing from the female's spinnerets at the tip of her abdomen and it can be there for only one purpose. The female is about to wrap the male up and eat him! Some sixth sense warns him of danger, and he leaps away in time, dangling on his safety line, then settling on a leaf. The female seems busy, back on the web, with some object she is carefully wrapping in silk and manipulating with her jaws.

'The instinct to mate is extraordinarily strong, and it brings the little male climbing back up his safety line and staggering back up the web. He looks a little unbalanced, but it's not from the shock of his experience, it's because he has only six legs left. The female got the other two when he sprang away from her, and this is what she is busily eating. . . . Seemingly none the worse for his grisly experience, he sits there running his remaining legs carefully through his chelicerae. Perhaps he is counting them. . . .'

The home in Catalpa Crescent says a great deal about its occupant. There are a great many books and piles of photographic equipment everywhere. One room is dominated by a large machine constructed by Densey's partner, Jim Frazier, to help them photograph very small animals. It is basically a film camera mounted on the carriage of a lathe; a gyroscopic control plundered from an aircraft enables them to move their subjects through every plane. When they are working together on a film, Jim Frazier is the cameraman and Densey helps him with animals: 'I'm the one who prods and pokes them and persuades them to do the right thing while Jim's got his eye to the camera. I'm also the one who looks after the animals' welfare.'

There is evidence of this animal welfare work everywhere: pots filled with fresh leaves for the caterpillars and grasshoppers and

The netcasting spider, *Dinopis subrufa* weaves an intricate net of silk which she wraps firmly round her helpless prey.

other creatures appear in unexpected places all over the house. Densey is very slightly defensive about it: 'My house is extremely untidy—housework is the last thing that occurs to me. Occasionally I have a good old spring-clean and that's it. There just isn't enough time to do housework when you've got so much else which is worthwhile and fascinating and fun to do.'

Even the moderately house-proud might incline towards despair at the sight of Densey's kitchen. It certainly is not dirty but she admits it is a muddle. Bread for the birds is soaking in the sink; several small green caterpillars that have escaped from somewhere wander across the draining board. On the dresser amid a jumble of crockery are a number of screw-capped jars that appear to be a quarter filled with earth or compost. On closer examination they prove to have other occupants—spiders. Densey seizes two of the jars and invites us to help her to photograph the spiders.

At first her preparations are mystifying. She fills a plastic washing-up bowl with water and places a rock in the middle, like a little island, where the spider will be photographed. 'We don't really want this one scuttling round our feet. This is quite a nice little spider and I don't think you need to be

The funnel-web, *Atrax robustus*, is the world's most deadly spider (this is the male).

worried about it, but if anything does go wrong, you just dial 999, ask for an ambulance and tell them it's a funnel-web bite; they'll know what to do, they're used to it. . . . Right now, we have to induce this little lady to come out of her jar and sit nicely and quietly for us on top of this rock.

Densey pulls on a glove and fiddles in the jar with some long forceps. Finally, she manages to coax the spider out. Considering its fearsome reputation, the Sydney funnel-web is an unexceptional looking creature: dull black in colour and about one centimetre across its back; and yet the bite is deadly: 'This is probably the most venomous spider in the world. If they get you to hospital you might last a couple of hours. Not that they can do very much for you because there's no antidote against the bite. Twenty years of research have failed to identify the poison exactly.'*

The reason for the basin of water becomes abundantly clear as the spider attempts to

* Since our brief encounter with the funnel-web, we have learned that after those long years of research at the Commonwealth Serum Laboratory in Melbourne, an anti-venine has at last been discovered. It works for monkeys but, as far as we know, has not been tried on humans. Presumably its value will not be known until some unfortunate is bitten.

Fortunately for Sydney's suburbanites, the female funnel-web spider is reclusive by nature. The number of human death's recorded from spider bite – red-back and funnel-web – is only 25.

make its escape and drops into the moat. Densey explains that accidents happen when funnel-webs fall into swimming pools, since although they look dead, they are in fact semiconscious and come back to life when fished out of the water.

Densey feels very strongly that spiders like snakes and toads, bats and cockroaches, are among the most maligned and misunderstood of all animals. 'There have only been about 13 or 14 recorded deaths (from funnel-web bites) and when you think of all the carnage on the roads, it's a bit silly to get upset about spiders. . . . People are inordinately scared of these spiders,' she says as she clicks away with her camera and prods the creature with her forceps 'and I think it's a very primitive reaction'. Among male humans spiders enjoy a particularly bad reputation because (according to Densey) men believe that male spiders are eaten by their mates. 'If only in the interest of female solidarity, let me set the record straight. Male spiders are not *always* eaten by their consorts! When they are, it is usually due to a breakdown in communications.'

But spiders are not perhaps as ill-regarded

as Densey thinks, for they frequently made honourable appearances in ancient myths and heroic legends. Not only did Robert Bruce learn his lesson in perseverance from watching the determined efforts of the web-weaver but Frederick the Great actually owed his life to a spider falling in his drinking goblet. The story tells how the King called for a fresh drink and a little later heard a gunshot ring out: his steward has been bribed to serve him a poisoned draught and believing he had been betrayed, shot himself. According to tradition, the ceiling of the Palace of Sans Souci was painted with a spider design in remembrance of this event. And yet spiders are not always – or even usually – the story-book heroes; it is not only Miss Muffet who has reacted to their presence in a less than positive manner. Mention of Miss Muffett has an explosive effect on Densey. 'Miss Muffett was the daughter of an entomologist, and far from being frightened off her tuffet by that spider, she probably collected it for her papa. The story was just put about by some male chauvinist of the time.'

Spiders certainly need their allies around Sydney and Densey feels that the paranoia about the funnel-web is deliberately cultivated: 'We don't have any large, fierce predators in this country and I think people feel this to be a shortcoming. A big, wild country should have big, wild animals, so they make the most of what they've got.' In Densey's view the only animal to be feared in Australia is man. 'The commercial exterminators play up the danger; they spray under people's houses and around their gardens and kill everything alive. It totally upsets the balance of nature.'

Densey frequently makes journeys into the outback alone; she once drove herself right around Australia and has twice driven across it. 'I had a van which I slept in at night and I was entirely on my own. People call it a hard and desolate country; to me Australia is a very gentle place. There is nothing in the land itself that frightens me, only people. I must admit that some nights when I was camped hundreds of miles from anywhere, by the side of some dark road and I heard a car coming in the distance I was petrified. But I would wander about the bush with a torch and not be afraid of anything. There is nothing to be afraid of.'

Although she is the easiest and most affable of companions, Densey Clyne claims, like many other naturalists, that she has 'never been as much at ease with people as I am with animals, even with spiders'. Densey recalls the day she brought a large tropical bird-eating spider from North Queensland and installed it in the house. 'The spider's name was Lulu and when Lulu came into the house, that was the day my husband walked out.' She admits that 'he's my ex-husband now. He doesn't like spiders so I had to make my choice, him or Lulu. Not that *he* put it like that, he's much too courteous. He just left quietly and took up residence in a high-rise block of apartments where there's not even a daddy-long-legs; you see, he likes nature through a plate-glass window, and I like it in contact, in touch.'

To be in touch with nature is so important to Densey that it has now become her whole life. Night and day she spends observing, photographing and writing about the lives of the inhabitants of her garden jungle. Like many naturalists, she finds that photography is her most precious tool 'because it's a way of stopping time. The lives of insects are so brief, and you might only see something happening once in your lifetime – something that no one

else has ever seen – but now you can collect that happening in a way that naturalists never could in the past. Collecting specimens is really going out, thank goodness. Who wants a collection of dead insects? It's as bad as hanging dead animals around your neck; the best possible way of recording them is with pictures and with words.'

As darkness falls in Turramurra, Densey ventures once again into her garden, this time armed with a floodlight on a lengthy cable 'to see what the night-shift is doing, because something very different goes on out here every night'. She is particularly anxious to show us the mating of the leopard slug (*Limax maximus*). As we cross the terrace we are warned to be careful where we put our feet. 'The slugs come here to feed on bird droppings and fallen leaves; they should be welcomed in any garden because they're scavengers, part of nature's vast army of cleaners and moppers-up, and they never feed on living plants.'

Coscinoscera hercules. The word 'caterpillar' comes from an old French word *chatepolose*, meaning 'hairy cat'; some are spiny as well as hairy.

Very soon we come across a pair of large, mottled creatures gliding sluggishly across the terrace, head to tail. This is the beginning of the courtship and we must be patient because slugs, like Englishmen, need time.

'Leopard slugs are beautiful: I mean one would never think about leopard slugs as being passionate or sexy or in any way lovely; mostly, people are horrified by slugs, but these are rather special because they have a very interesting sex life.' Densey's commentary issues softly from the darkness. The two slugs negotiate the terrace, still in a head-to-tail position and begin to ascend a tree.

Whereas most slugs do it modestly under a stone, leopard slugs mate in the open. Since slugs are hermaphrodites, Densey is intrigued to discover how the different sex roles are determined: 'each one is both male and female, but one seems to play a leader's role and the other a follower's role and nobody knows why.' After half an hour of very slow progress up the tree, the slugs begin the next stage of their courtship. They start to circle, still following each other, and laying down a mass of slime. We are all familiar with the silvery mucus trails laid down by slugs and snails, but this one is obviously not for roadmaking. According to Densey 'it will provide a swing for their trapeze act'. Soon, incredibly, the coupling of these laborious and least ethereal of creatures happens as they gyrate in mid-air, suspended on a silvery cord.

In her book, *The Garden Jungle*, Densey describes the final stages of that courtship: 'Now the circling slugs move side by side, the following one having caught up with its fellow by moving to the inside. They head vertically downwards over the edge of the sloping tree trunk, producing an obvious string of mucus from their hinder ends. Their bodies start to entwine. By the time they drop clear of their support, the two slugs are locked in a slippery embrace, held safe only by the rope of tough mucus as they writhe together in mid-air. Their sinuous movements rotate them with a gentle rhythm, first one way and then the other, as they slowly descend.

'Just to be different again, with slugs everything important happens at the front end. So from under the loose edge of the "mantle" that covers the front of their bodies, the dangling leopard slugs are now producing identical male organs. They are pearly white, flexible rods that extend fully and reach out for each other. They touch, then suddenly they corkscrew into swift conjugation, appearing almost to fuse together into a single structure that gleams bluish-white in the beam of the floodlight. It hangs below and between the slugs, linking their heads together, and it continually changes shape. Now it has grown a translucent spiral flange with sinuous edges that flare and expand, only to be absorbed again into the main structure. It is within this mobile, sculptured receptacle that the essential exchange of sperms takes place at some stage of the slugs' aerial gymnastics.

'At the end of the performance, which can last up to half an hour, the slugs will part quite swiftly, the mating organs pulling free of one another and retracting into a hole in the side of each head, while the bodies reluctantly disengage. One slug usually slides down the other and drops to the ground. The other descends the mucus rope, sometimes eating it, and takes the long way back to earth. In a few days these two slugs will lay their fertilized eggs under a stone or in a crevice in the soil.'

Such powers of observation and awareness are typical of Densey Clyne. In another passage from *The Garden Jungle*, she describes the mating behaviour of the praying mantis in a masterpiece of acute observation and wit. 'It is common knowledge that during the mating season male mantises tend to lose their heads—and I mean it literally. If a lusty male's approach to a likely female in any way lacks finesse, praying will do him no good at all. In fact he might soon find himself without any front legs to pray with, as they disappear like sticks of celery between the jaws of his intended. There is nothing personal in the female's attack; she just has this habit of regarding anything that moves near her as fair game, which makes sex something of a health hazard for male mantises.

'Because of the cannibalistic tendencies of female mantises, courting males try a variety of approaches. Like this one. On the tip of a leafy stem among the bushes a green mantis sits, her eyes set like huge red rubies at the upper corners of her small, triangular face. Gazing mildly into space she waits for whatever the night may bring. The night has, in fact, already brought a male of her species to a nearby leaf. He is out of range of her eyes, and judging by the hesitant nature of his approach it is not part of his courtship procedure to try and turn her head.

'This male is slightly smaller than the female, more slender of form, equally bright of eye. Male mantises are timid creatures, not easily handled. Is it my imagination that this one has a frightened expression in his bulging eyes? He looks ready to make a quick getaway at the slightest movement on the female's part. Fifteen centimetres away from her and still unobserved, his feelings get the better of him – he decides to take the female by surprise. His green wing-covers part, he unfurls his delicate, transparent wings, and takes a graceful flying leap towards her. But now her head is turned by his attentions and as he misses his mark and ends up clinging by his front feet to the rim of her leaf, that mobile triangle swivels down and around. For a second or two she gazes into his eyes, her dangerous raptorial (or grasping) legs still folded. He is immobile, paralysed by her proximity. Then like a flick-knife her legs open and lunge at him. The action is too swift

The improbable aerial mating of leopard slugs, *Limax maximus*. The blue-white extended sex organs appear at the front of the body.

for the eye to follow. The reaction is equally swift. The male drops away as if shot.

'Sexually aroused males have short memories for rebuffs, and this one comes back to try a more cautious approach. One leg at a time, millimetre by millemetre, he moves towards her, crouching low. She, body held high to mark his approach, keeps an interested eye on him. He keeps a wary one on her. His strategy pays off. Caution is obviously the watchword, and his respectful approach leads to acceptance. Now he rides the female's back, his front legs clasping her just behind the front thighs. It is these weaponlike front legs of hers he'll need to watch, but he'll be safe as long as he keeps his embrace tight and his head down. Especially the latter, because if the hunting hasn't been good and if she happens to look back over her shoulders and see him – well a mantis's memory is short; she can easily reach around with her raptorial legs to seize her lover's head, and he will die, enraptured, as it were, by her embrace.

Densey Clyne has the unusual ability to see and to describe the small beauties of the natural world around her. She can find drama of all kinds, romance, comedy and tragedy, where most of us find nothing at all. Like most of our naturalists she does not regard anthropomorphism as a sin; she knows that if we cannot understand animal behaviour, however extraordinary, in human terms, we cannot understand it at all.

'The ways of these creatures may not always be our ways. But the lives they live are as bizarre and fascinating as any story of science fiction. I think they are as worthy of preservation as any of the world's wildlife. And they can be found on a short safari along anyone's garden path.'

Reference

CLYNE, Densey, *Wild Flowers of the Outback*, Rigby, 1973.

CLYNE, Densey, *A Guide to Australian Spiders, their collection and identification*, Nelson (Australia), 1977.

CLYNE, Densey, *How to Keep Insects as Pets*, Angus and Robertson, 1978.

CLYNE, Densey, *The Garden Jungle*, Collins (Australia), 1978.

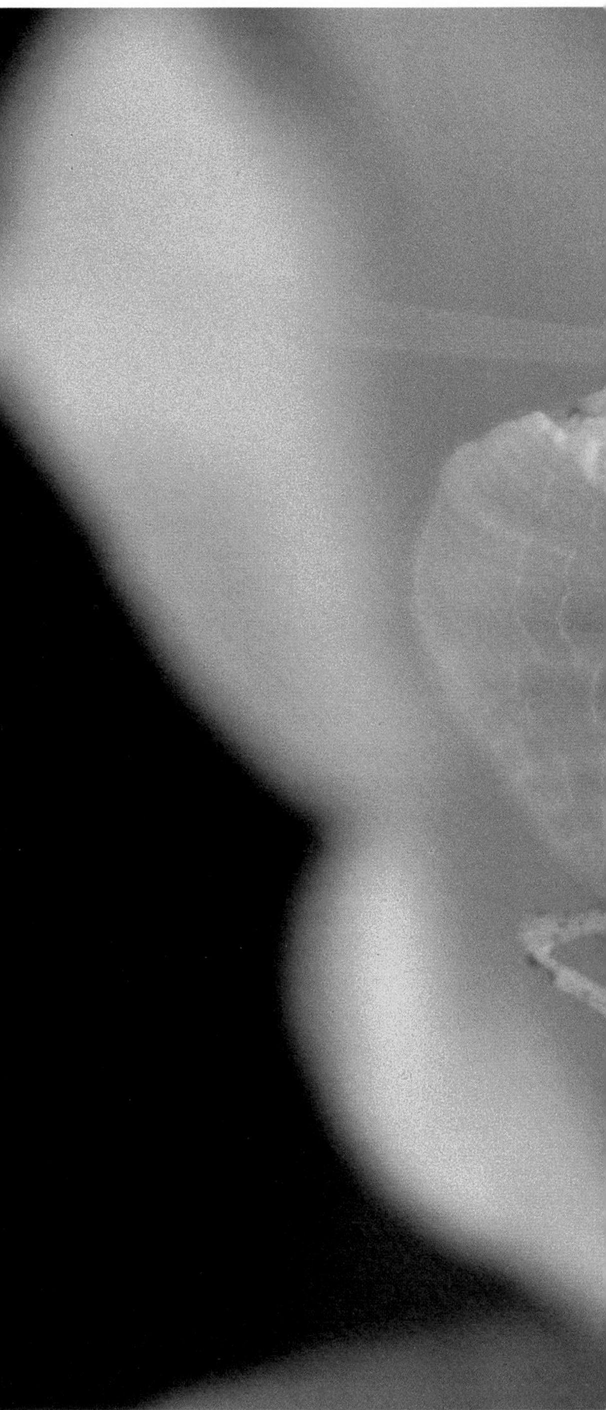

A supremely effective predator, the mantis is well camouflaged for lying in wait among leaves, ready to pounce on unsuspecting prey.

KONRAD LORENZ

'What am I? A Watcher of Animals!'

Altenburg and Grünau, Austria

Konrad Lorenz, aged 77, energetically enjoying the autumn of his life, is proof positive of that most attractive feature of a love of nature – retirement is not to be feared but welcomed.

Officially, Lorenz retired several years ago after a busy working life of teaching, writing and research during which he contributed as much as any man alive to the science of animal observation. With Niko Tinbergen, he is regarded as the father of ethology, the science of animal behaviour, a status officially recognized by a shared Nobel Prize in 1973.

But when the time came for Lorenz to give up a full-time teaching role he simply moved home to the baroque villa built by his father alongside the Danube and used this reward of spare time to do what he does best – to stare at animals. In particular to stare at fish: Lorenz in retirement simply settled in front of what must be the largest private tropical fish tank in the world; a 9-foot cube of tepid brine protected by huge, throbbing pumps and heated by blinding mercury vapour lights. For four hours of every day, Lorenz lounges in an armchair staring at his fish. Caught in this pose he could be mistaken for an old man gently nodding away his old age. But such an impression is totally wrong: bright eyes flicker all the time and every so often the big

Konrad Lorenz, at 77, with greylag goose.

The favourite birds of Lorenz's studies, two graceful greylag geese in flight.

head with its shock of white hair will lift suddenly and Lorenz will cry 'Ah—there!' even though he is quite alone.

Lorenz is still very much at work. The darting movements of the fish, which to anyone else are no more than a colourful kaleidoscope, still contain for Lorenz innuendoes of behaviour and vital messages about their essential nature. After each morning and evening session, Lorenz drags himself away reluctantly, for there are books to be written, scientific papers to be edited and reviewed and an avalanche of mail that can never quite be managed. But in the grand study lined with books and files there is also a model electric train. It takes up a good half of his desk. Lorenz took two weeks off to build it, sculpturing mountains out of papier mache. 'Perhaps you shouldn't mention the train set', he smiles, 'there are enough people think I'm an old damned fool already'. But the fact that the man can play with a toy train is as important a pointer to the real nature of Konrad Lorenz as may be found in his eminent books like *On Aggression* or *King Solomon's Ring*.

A strong iconoclastic streak runs through this Austrian. He rejects totally man's more pompous view of himself and believes vehemently that a sense of humour and the ability to play are gifts that only dolphins and the very advanced primates and hominids have developed. 'Humour is a very serious thing.' Lorenz says, then winks to confirm he is well aware of the paradox. More seriously: 'Humour is part of self-observation and self-discipline. A man who has some real humour can never become a megalomaniac. For example, if you had seen Hitler and Mussolini talking – their mouth and facial expressions – you would have realized instantly that these persons were completely devoid of humour.' Humour, in the form of self-observation, Lorenz constantly applies to himself. He adores limericks, even bad ones.

Asked, very seriously, how his interest in animals began this *eminence grise* of ethology grins and replies: 'I wanted to be an owl, because owls are not put to bed early. Then I found out that owls could not swim, so I decided to be a duck. I got a duckling newly hatched which my mother brought me. My father was against it because he was against cruelty to animals and he thought I was going to kill the duckling. But I didn't: I raised it. I became infected with "duckopholy".' More on this later because 'duckopholy' is not the joke phrase it may seem. It is Lorenz humourizing one of his greatest discoveries.

In fact, Lorenz is probably incapable of recalling the exact age at which he became

Above: The study of fish (in Lorenz's own aquaria) is a favourite pastime.

Right: The colourful grace of a mallard duck in flight.

fascinated by animals; the stories of his infancy indicate he might even have crawled in search of them. Certainly he remembers with affection his nurse, Resi Führinger, who had what he calls 'green fingers with animals'. The very young Lorenz collected river fish and crustacea, and his otherwise Victorian parents showed remarkable tolerance towards his often strange acquisitions, although at this time dogs were banned. Lorenz's father, a surgeon of international repute, was concerned that dogs might carry microbes dangerous to the health of his infant son. So instead, Konrad had a crocodile for a while, and when that had to go because conditions at Altenberg proved too cold, the first dog – microbes and all – was wheedled into the house.

There were also birds of all shapes and sizes, including ravens and cockatoos, and even a Madagascar lemur – a not insubstantial ape – that leapt around the house in wild abandon confiscating cigarettes from visitors.

What the neighbours thought about all this is hard to say; the Lorenz house has a tall iron fence around it. But there is a story, which is still told in the village, of the Devil, complete with horns, claws and a tail, appearing among the high turrets of the house, attended by a flight of screaming black birds.

It was Lorenz in party costume, but it was not a jape. As usual he was pursuing his animal studies. By this time Lorenz was a recognized researcher and had published a noteworthy paper on jackdaws from observation of a large colony that he had established on the roofs at Altenberg. As usual his relationship with his subjects was intimate, but Lorenz also knew that his jackdaws would attack any black object intruding on their colony. Thus, when he needed to ring fledg-

lings he would don his devil party costume and suffer the attacks of the adult jackdaws content in the knowledge that his personal friendship was protected by the disguise.

All this time at Altenberg there were flocks of ducks – but then ducks to a man who admits to being infected with duckopholy, have always been very special. It is time we considered this strange disease.

Two hours' drive from Altenberg, on the edge of a wildlife park in the beautiful mountain resort of Grünau, Lorenz is still actively involved in the study of duckopholy, although the fieldwork is now done by his assistant, Sybille Kalas.

The birds they study are not in fact ducks but geese – exquisite pink and silver greylags. In the perfect peace of these snow-capped mountains, Lorenz and Kalas are mother and father to generations of hand-reared greylags. This peculiar relationship is the product of 'imprinting'. In the 1930s Lorenz discovered that newly hatched goslings follow any moving object they can see, be it a mother bird, a man, a rubber balloon or even a cardboard box! Having once followed a particular object they will prefer that object to others and after a while will follow no other.

Imprinting, or attachment behaviour as it has been called, fell like a bombshell on the natural science establishment. Until then most scientists, with Freud in the van, had believed that the mother met a young animal's psychological need for food and warmth – the 'cupboard love' theory of attachment – and that the young knew by experience that the mother satisfied all their needs. The Lorenz

Grünau, Austria. In this idyllic setting, Konrad Lorenz and his assistant Sybille Kalas observe families of greylag geese.

A sublime study in pink. silver, black and grey; it is not difficult to see why Konrad Lorenz so admires the proud greylag.

findings knocked deep, painful dents into this and two other sacred beliefs: motherhood per se, and the theory that you could make 'assumptions' about animal behaviour. Lorenz proved that all behaviour had to be observed or established by experiment. Scientists of the day (and possibly a few mothers) did not take too kindly to this assault, but Lorenz's experiments were soon repeated and verified.

As the years went by it was recognized that there was substantial evidence to indicate that attachment behaviour also developed in a comparable manner in mammals, including man; more simply, that much of our behavour was the byproduct of 'instinct'. This was a revelation that man – or quite a lot of the human race – found very hard to stomach. Lorenz incurred what he himself described as 'the fanatical hostility of all those doctrinaires whose ideology had tabooed the recognition of this fact'. (*Studies in Animal and Human Behaviour* Lorenz 1970.) The irony of it was that Lorenz had also shown that such hostility was in itself instinctual: 'No other force could account even for the fury with which men hold to their beliefs, against all evidence and all reason, including the still prevailing belief that they are not instinctual.' Lorenz himself had no grounds for doubt. By then he was mother and father to a number of imprinted geese, and has remained so, very contentedly, ever since.

What is somewhat extraordinary is that Lorenz believes with some justification that he really discovered imprinting when he was five years old: 'I can still remember how my baby duckling made a "lost" sound and that I realized it was crying. I tried to console it by quacking like a mother duck.' (Lorenz gives a perfect imitation.) 'I remember as if it were yesterday the moment when this duckling ceased to weep and gave the greeting sound. And then I moved away from it, creeping on all fours backwards down on the tiled floor of the kitchen. I remember my great happiness when the duckling started to follow me.

'There was a little girl living nearby and she had a duckling as well.' (That little girl is now Lorenz's wife, Gretl.) 'We played at being mother ducks all through the summer. But my duckling was newly hatched and hers was a day and a half old. Now her duckling never followed as well as mine because its imprinting age was past.'

But the ramifications of imprinting extend much further than baby ducklings becoming maternally bonded to humans. Lorenz believes the process works both ways: 'My wife is three years older than me and when we were those small children playing with ducks, I got imprinted on *ducks*, but my wife didn't. I am now quite sure you have imprinting-like processes at the root of the likes, dislikes and interests in man. I still feel a certain peculiar attraction for waterfowl of any kind.'

Imprinted birds showing Lorenz 'the loyal attachment they would display to their real parents under natural conditions'.

Imprinting on humans does not imply a lack of 'normal' maternal instinct; on the contrary, geese take great care of their offspring.

Imprinting is, of course, just one of the many innate behaviour patterns Lorenz has identified during half a century of searching. It is also less controversial than some of the others, for the man has had more than his share of controversy.

The main theme running through all of Lorenz's work is essentially an idea that tends to stick in the human gullet. Stated at its simplest it is that man is just another animal, and like other animals we humans are born with important behaviour patterns. We do not learn everything in some sublime special school of humanity; some of our more poisonous characteristics, such as aggression patterns, are inside us inexorably, as they are inside and impel Lorenz's tropical fish or the greylag geese.

Why such a theme should continue to be received with such hostility is hard to understand. There was some excuse when Darwin drew the wrath of man-worshippers with his *On the Origin of Species*, 120 years ago; the fact that it is hardly diminished today is an indication of how fiercely man clings to his belief in a special estate. Lorenz, as we have already pointed out, is an iconoclast with an almost evangelical streak when it comes to fixed theories: 'It is a very good morning exercise for a research scientist to discard a pet hypothesis every day before breakfast. It keeps him young.' It has certainly kept Lorenz mentally young, but it is difficult to see how the principle could be generally applied. After all, there are not that many scientists with theories enough to discard one a day. With Lorenz it is a different story: he has written a library of theories, and most of them are more statement than theory – that a particular aspect of behaviour is true because he has carefully observed it (and others have checked it).

Perhaps Lorenz's most dramatic collection of concepts of this calibre was published in arguably his most famous book, *On Aggression*; of these concepts, one of the most startling is that there is no such thing as a 'wild beast'. In a sense the book is mistitled. The more you read about aggression, the more Lorenz convinces you of its role, its logic and its point; a meaning far removed from the mindless connotation we so often give to the word.

He does not believe, for example, that there is any kind of 'extermination streak' in nature where vital interests are not in competition. And where eater and eaten are going about the business of survival, the eaters never carry their predation to the point of extinction; a state of equilibrium is always established between them, endured by both. In fact, extermination – mindless killing, if you like – exists only where there is competition, with too many parallel species eating the same thing.

It all sounds very 'non-aggressive', and in his book Lorenz gives a number of examples which any layman can observe. Like the way a hunting dog obviously enjoys the hunt and is no more 'aggressive' towards the rabbit than a butcher is 'aggressive' towards the chickens he slaughters for our suppers. Then there is the 'mobbing' behaviour employed by groups of animals to drive off a predator, which, Lorenz points out, actually has the educational value of showing young animals what a predator looks like.

Perhaps the aggressive pattern that most closely approaches the 'savage beast' concept is what is called 'critical reaction' – a fear-dominated 'cornered rat' behaviour when an animal will fight to the death with almost 'mindless' ferocity. But in these instances, the animal has no option and knows that in this rare circumstance it must throw everything into the fight if it is to stand a chance of survival.

'Mindless' violence has been given several names by man. One of them is 'evil'. But Lorenz dislikes both the blind term and even more so the metaphysical concept of evil in terms of aggressive behaviour. To avoid the ambiguity of such terms, science calls what appears to be unthinking, almost natural aggression between animals of the same species *intraspecific* aggression. And again, Lorenz stresses that it is definitely not pointless. If a large group of animals of one type and thus the same feeding pattern were to gather and stay in the same spot, they would soon eat their way to extinction. Intraspecific aggression then, where similar animals are aggressive towards each other, has the effect of spreading the group as widely as possible over the available space, or ecosystem.

So much for 'animal' aggression, but Lorenz is not at all convinced that what might be called 'mindless violence' has been avoided

Greylags form pair bonds that can last their lifetime, and are fiercely protective of their mates during courtship. Although fights occur, aggression is usually ritualized with competitors taking the hint before actual violence ensues. Winners are not above demonstrating their psychological victories.

by man. 'The aggression drive [in man] has become derailed under conditions of civilization' he wrote. It brought the now traditional howls of protest from the human establishment. For Lorenz it also revived the accusation that his work supported racial elitism, even though his defence of man's genuinely individual characteristics, such as humour and conceptual thought, has been as vehement as his insistence that we must, in the interests of our own sanity, recognize our innate characteristics. 'Man is so proud of his great capacity for conceptual thought, he likes to think that his behaviour is governed by rational, conceptual thinking. He is offended if you tell him that when he just jumped into the water and saved the little girl, he did so by programmed behaviour and that a chimp would have done the same.

'He is ashamed of having poor relatives – that is the way I would put it. But the evolution of conceptual thought in man brought with it responsibilities, because man could then *foresee*. He could examine himself and foresee the consequences of his actions. And this makes man very different from animals. But this does not change the basic foundation of all instinctual activities, of the whole mechanism of perception. Everything that makes a higher animal is still in man. But on top of this there is conceptual thought with all its glorious consequences and all the horrible dangers which are incurred by the freedom and manipulation of *instinctive* man.'

Understandably, Lorenz now has considerable sympathy for Darwin; 'I think I am acting as a spear – a lance – going in front of Darwin; going into regions Darwin had not penetrated.' Then he smiles and admits: 'There's nothing new in what I say. There's nothing new in my whole sermon on humility and so

on. It's addressed to people who do not yet understand Darwin.' This is definitely too modest. Lorenz understands Darwin as well as any man alive and he knows full well that he stopped well short of actually saying that man is descended from animals. Darwin implied it; Lorenz shouts it, it is his message to the world. He believes that man, the evolving animal, subject to natural selection like every other animal, is running himself down a lethal blind alley.

Such blind alleys exist quite commonly in nature and result in what Lorenz calls 'bizarre' selections. He points to the excessive antler structure of stags. As antlers were used to assist rival fights, the bigger they were the better the stag's chances of success – up to a point. But natural selection took the characteristic to extremes and now stags use their sharp forehoofs for defence rather than their ungainly antlers. Indeed a stag weighed down with this unweildy head gear is obviously more vulnerable than a stag without it.

Lorenz's teacher, Oscar Heinroth, had similar feelings about the cock argus pheasant, which has secondary wing feathers so huge that the bird can scarcely fly. Hen pheasants are sexually stimulated by these huge display feathers and, as a result, the cocks with the most ludicrous display inevitably father the most progeny. So cock argus pheasants selectively evolve bigger secondary wing feathers, despite the fact that their development is making the bird ever more vulnerable to predators.

Lorenz records that Heinroth once told him, jokingly, 'Next to the wings of the argus pheasant, the hectic life of Western civilized man is the most stupid product of intraspecific selection!' But Lorenz, who agrees, no longer thinks of this as a joke. He is convinced that man has 'precipitated' himself into a melée of industrialization and commercialism, suffering as a result 'managerial diseases' and 'torturing neuroses'. 'They succumb to barbarism because they have no more time for cultural interests,' he expounds. In Lorenz's view this workaholic state of Western society is as bizarre a development as the argus pheasant's ludicrous wing feathers, and just as silly because, as Lorenz points out with ethologist logic, man could easily agree to take things more quietly. He also keeps insisting, to the fury of the protagonists of man, that we do not have the excuses that may be accorded stags or argus pheasants. They are primitive animals with little or no understanding of their basic drives nor the ability to control them, whereas 'unlike any creature before him [man] has mastered all the hostile powers in his environment', Lorenz states uncompromisingly.

Lorenz is so concerned about this aspect of man he is even prepared to use the word 'evil', rejected in application to other animals, to describe extreme forms of human aggression, and he has worked out a theory to show how it happened.

By the time of the Stone Age, man had essentially separated himself from the basic imperatives of the animal kingdom. Weapons, clothing and social organization set him free of the primary life requirements: food, warmth and the threat of predators. Aggressive drives in man were turned into intraspecific arenas – wars between men, rather than between man and nature. For tens of thousands of years human selection was shaped by these tribal wars leading, Lorenz believes, to 'evil intraspecific selection' that must have evolved as an extreme form of the so-called 'warrior virtues'. 'Which unfortunately', Lorenz warns, 'many people still regard as desirable ideals'.

Lorenz says he is addressing himself to people who do not yet understand Darwin, but it would be more accurate to interpret his work as a crusade to get people to face up to the implications of Darwin's theories.

Because he still lives in an age of neo-Darwinism and because he has been so insistent (and for the most part right) about man the animal, Lorenz continues to suffer something of a reputation as a human denigrator. In fact he patently has a deep, subjective love of all animals and is only in conflict with the protagonists of man because he refuses to see man as a totally separate and unique animal.

He does not, for example, think that there is anything like humour in non-human animals, other perhaps than in chimps and dolphins. On the other hand, man's humour is not in Lorenz's opinion a unique 'gift' but something that we have acquired as our brains grew larger and more complex. He regards it as a very modern development of our culture, evolving through the centuries, and because such things as culture and humour are evolutionary, Lorenz speculates intriguingly upon the possibility of advanced man being able to teach the lower animals. 'If you see a chimp in the Gombe Stream Reserve freezing and suffering in the tropical rains, and if you watch them building nests in the trees, you wonder why those silly beasts don't build a nest and stand *under* it.

'One might teach a chimp to build a roof and see whether his free-living fellow members of the species learn to imitate him. The power of imitation is there, the advantages of a roof are obvious. They would like to stand under a roof in rain. And it would be worth while because that would be the first rudiment of culture.'

One of the great joys of listening to Lorenz when he talks of animals is the realization that he has the ability to think like, or for, an animal. He is concerned about shivering chimps, whether his geese are happy or sad, the mental state of his dogs. Most people who have followed Lorenz's long career are aware of this intense private, as opposed to academic, interest in animals. For them it came as no surprise when, in 1954, the guru of advanced ethology published *Man Meets Dog*; others, however, were very considerably startled.

It is not the popularist, pet-lover pocket book the title might indicate. But it is a million miles removed from the hundreds of scientific papers and manuscripts that bear the name Lorenz and often require a specialist dictionary to be read. It was, quite simply, dedicated to 'all those who love and understand dogs and cats alike' and it opens with what could be called a fairy story: Lorenz's rather romantic concept of how dogs became domesticated.

'The knowing old hunter who was their leader lost his life a few weeks earlier; he was wounded by a sabre-toothed tiger who tried one night to steal a young girl from the band . . .' It is a delightful mixture of Brothers Grimm and science-fiction, sugar-coating some of Lorenz's ethological theories. A few paragraphs later, for example, a new leader takes over the band. He has less muscle power than the former leader 'but his eyes are brighter and his forehead higher and more arched'. (Darwinian natural selection has been at work!) In Lorenz's story, this evolving man comes to terms with the jackal packs that follow the human band, sharing their kills and eventually training compromised jackals to track and hunt for the humans. It is as good a theory as any and it is important in one aspect – it gives credence to the peculiar symbiotic relationship that exists between man and dogs. Very early on in *Man Meets Dog*, Lorenz has identified this extraordinary relationship. 'Only two animals have entered the human household otherwise than as prisoners and become domesticated by other means than those of enforced servitude; the dog and the cat.' He accepts that cats have also kept at least one paw in the wild state. So in terms of a total involvement with man, the dog is unique in the animal kingdom.

In Lorenz's world dogs are unique – the only animals to have entered the human household freely, not as prisoners.

Lorenz certainly thinks they are. And it gives him the opportunity to talk about characteristics which he patently values greatly, like fidelity. There are some wonderfully evocative descriptions of how Lorenz and his chow, Wolf, walk the countryside, rebuilding a relationship that has been truncated by the war years when Lorenz was a prisoner in Russia. Lorenz becomes the chow's pack leader, while building a respect for the dog which, more than anything else, demonstrates why Lorenz is such a good ethologist. He loves animals and happily enters the dog's world where *he* must earn respect and status: 'It is extremely funny how everyone – myself included – who knows this proud, imperious dog feels flattered if he honours them with a majestic indication of his favour.'

If all animal scientists worked from this standpoint, subject only to the kind of scientific discipline which Lorenz's conclusions (rather than his observations) demonstrate, we would learn a great deal more. *Man Meets Dog* is probably the best example yet of how it is possible to combine anthropomorphism with a meaningful search for knowledge. Conservatist scientists have always suggested that the anthropomorphic view is dangerous and confusing, but the price they pay is observation often so remote that it is quite meaningless.

A polyglot mob of dogs guards Lorenz at Altenburg; all are mongrels, many with a bit of chow or, as Lorenz puts it, 'Wolf', in their lineage. Lorenz is openly anthropomorphic about his dogs and even thinks the anthropomorphic view might be the best one. Certainly it is the way to think if you wish to avoid causing the dogs any suffering: 'I tell animal-loving audiences, "your dog is much more stupid than you realize but emotionally he is much more similar to you than you realize, and this is important from the point of view of not committing cruelty to animals. For example, I'm sitting on the Dover boat and I'm suffering horribly from sea sickness. I should like to die at that moment. But at the same time I'm not all that sad because I can see the breakwater and know that in 30 seconds my suffering will be over. That's beyond the capacity of a dog. If you leave your dog at home and go away for a few days, the suffering of the dog is exactly the same as if you would have died, because he cannot

The mating and courtship of the greylags is a formal ritual.

The swift act of mating seems brutal with the female held down firmly by the neck.

Mating takes place on water, preceded by the gander dipping his head and wetting his back.

The gander holds his head in a proud 'S', with both wings slightly raised.

But the goose is a willing part and is soon swimming again.

The gander displays to announce a successful pairing.

foresee rationally that you are coming back".' It seems so obvious but how many humans, even pet lovers, can see this vision of the world, their pet's point of view?

And for that matter how many scientists? Lorenz holds firm opinions: 'If you ask a man "what is science?", the usual answer will be physics and chemistry based on analytical mathematics. It is really a mental illness of humanity to believe that something that cannot be defined or described in terms of the exact nature of science, or cannot be verified by analytical mathematics, has no real existence. In other words, our values are emotional: friendship and truth are illusions. If you believe *that* everything loses its sense. That is not science because science is to do with the whole man. To cut off the subjective, emotional side of humans is a dirty lie.'

This animal-eye view of the world is certainly one of Lorenz's most attractive characteristics and it may also be indirectly responsible for his unique insights into natural science. At one time he thought his contribution to how we think, epistomology, might be his most important work.

But over the last decade he has become very worried about the human race. He wrote a complete book on the subject *Civilised Man's Eight Deadly Sins*, of which the most deadly was what he termed the 'genetic degeneration of humanity': adverse changes to man's basic structures as a result of domestication, examples of which can be seen more graphically in overweight beef cattle or unwieldy domestic geese and hens.

Since his retirement, however, Lorenz's fears for humanity have become more specific and immediate. 'It is to be feared that humanity will destroy itself long before it has had time to degenerate.' He sees a 'headlong avalanche in the destruction of our habitat and our ecosystem', and shares with activists a third his age deep concern about the way we are using up our energy reserves. 'I have said a hundred times and I do not mind repeating it a thousand times that a child in high school that has understood the principles of compound interest must understand that exponential growth in finite space must lead to catastrophy. The realization of the danger fortunately also grows in an exponential curve and the only question is whether these two curves will intersect before the point of no return is reached. That point may be very, very near.'

Such doomsday theories are not new; they are even fairly common in men of advanced years. But Konrad Lorenz is no ordinary man and on his lips it has a deadly ring. He has better insights than most into the nature of things and when such a warning is expressed by the world's most eminent ethologist, we would do well to listen.

Oppressive thoughts of this calibre might also have produced an ancient cynic and a deep sense of futility, but fortunately they have not. He has defined the reasons for us being in so sorry a mess and can offer a way out: 'Far too much of civilized mankind today is alienated from nature. Most people seldom encounter anything but lifeless, manmade things in their daily lives and have lost the capacity to understand living things or to interact with them. That loss helps to explain why mankind as a whole exhibits such vandalism towards the living world of nature that surrounds us and makes our way of life possible. It is an important and worthy undertaking to try to restore the lost contact between human beings and the other living organisms of our planet. In the final analysis,

The nature reserve at Grünau provides sanctuary for a variety of ducks and geese, including these sleepy barnacle geese.

the success or failure of such a venture will determine whether or not mankind destroys itself along with all the other living beings on earth.' (*The Year of the Greylag Goose.*)

That statement contains the reason why Lorenz is not the cynic he might be expected to be: he still has his dogs, his birds and his fish in their tanks; his encounters with nature and living beings go on today and will continue to the end of his days. He is still at best a watcher of animals and that, as he has defined in the introduction to this book, means he will stay a 'happy researcher'.

Which is all very well for Konrad Lorenz, a sane man whose old age will see him merging ever deeper into a fabric of nature that most people would regard as close to paradise. He is almost recycling himself. With great authority and immense scholarship, Lorenz has written on the wall of so-called civilized society and now, like the ethereal subject of that original poem, is about to move on. In a sense he is lucky. There are still wild geese migrating annually to Grünau, wolves still run in wild families through the northern forests of Europe and the American continent, and the wildebeeste of the Serengeti can still just make the annual walk upon which their continued existence depends.

Lorenz can still be a prophet. But his warning is that the rest of us may have to settle for postmortems. If Lorenz is to be believed, the clock of life on earth is set at ten minutes to midnight and that reality changes the role of the naturalist and the nature scientist. We are well past the time when an interest in nature was a suitable activity for English clerics or retired school mistresses.

It may well be that Lorenz's science of ethology is the science of survival and that the men and women featured in this book, the game wardens, bird watchers, entomologists, botanists, biologists and the Sunday walker who takes care not to step on a cowslip, are part of an emerging community which has the survival of all livings things, and as such this planet, in common.

It seems to us no accident that natural history, ecological awareness, animal conservation and concern for the environment are blossoming interests in contemporary society. They are not, as some have suggested, easy political alternatives distracting the public from more intractable issues.

Konrad Lorenz and with him, the other naturalists featured in this book have shown unequivocably that everything must pale against the threat of the naked ape running amok in his one and only garden.

Reference

LORENZ, K., *King Solomon's Ring*, Methuen, 1952.
LORENZ, K., *Man Meets Dog*, Methuen, 1954.
LORENZ, K., *On Aggression*, Methuen, 1960.
LORENZ, K., *Studies in Animal and Human Behaviour*, Methuen, 1970.
LORENZ, K., *Civilised Man's Eight Deadly Sins*, Methuen, 1974.
LORENZ, K., *The Year of the Greylag Goose*, Eyre Methuen, 1979.
NISBITT, A., *Konrad Lorenz*, Dent, 1976.

A sapphire sky and the tranquility of Grünau – a perfect setting for Lorenz's study of his beloved geese.

Authors' Acknowledgement

The book *Nature Watch* and the television series of the same name have been more than two years in the making, and a very large team has been involved in the preparation of the material for the various media. We simply do not have room to mention many of the people who gave us vital assistance (like Johnny X who poled our punt through the Okeefenoke swamp or Johnny Y who performed the same function in Papua New Guinea). We thank you all.

Without the following, this book would never have happened:

Our editor Val Noel-Finch; designer Jeanette Graham; picture researcher Heather Lowe; our consultants: Bernard le Gette in America and Dion Bohme in Australia; researchers Geoff Raison, Tony Budd and Bradley Borum; producers Colin Luke and Ashley Bruce; film editors Kit Davies and Paul Cleary and their assistants; and cameramen Peter Greenhalgh, Gerry Pincus, Ian Hollands, Richard Kemp and Charlie Lagus; Bob Halstead for his assistance with the Papua New Guinea experience; and last but by no means least the two Marys: Mary Crossley and Mary Richardson who against all odds allowed us to meet our deadlines.

The authors and publishers would also like to thank William Collins Pty Ltd for permission to quote from *The Garden Jungle* by Densey Clyne; and Eyre Methuen Ltd for permission to quote from *Man Meets Dog* and *The Year of the Greylag Goose* both by Konrad Lorenz.

Picture Credits

The publishers would like to thank David Warmington for the maps on pages 11, 27, 45, 59, 75, 87, 105, 121, 137, 153, 169 and 187; and the following for permission to use their photographs (numbers refer to page numbers):
ATV: 82, 106, 114, 169, 188, 199; BBC Hulton Picture Library: 22; Aquila (Edgar T. Jones): 74; Robin Brown: 81, 83, 87–89, 102, 113, 120, 125, 127; Ashley Bruce: 17, 23, 60–61, 73; Neville Coleman: 152–155, 157–165, 167; Cornell University: 90; Tom Eisner: 86, 92–101; Des Gershon (ATV): 10–16, 19, 45–46, 48–50, 53–54, 56, 63 (right); Liz Hetherington: endpapers, 20–21, 27, 37, 44, 65–68, 70–71, 136, 139, 141, 146–147, 149, 151, 190–192; Richard Kemp: 26, 32, 39, 186, 189, 196–197, 200–201, 203–205; Chris Knights: 77; Al Langston: 121–122, 129, 132–133; Margaret Mackay: 137; Roy D. Mackay: 142; Mantis Wildlife Films: 168, 171–175; 177–179, 181, 183, 184–185; Laura Medved: 105; Pat Morris: 143–145; LuRay Parker: 9, 124, 128, 131, 135; Radio Times: 187, 193–194; Lisa Ringer: 115; Lynn Rogers: 104, 107, 109–112, 117–119; John Tucker: 47, 51–52, 55; Bobby Tulloch: 28–31, 33–36, 40–43; University of Minnesota: 75, 79; Jimmy Walker: Title page, 6, 58, 62–63, 64, 72.

Index